Patric Pham

Hybrid-Kinematic Mechanisms

Patric Pham

Hybrid-Kinematic Mechanisms

Designed for Machine Tools

Südwestdeutscher Verlag für Hochschulschriften

Impressum/Imprint (nur für Deutschland/ only for Germany)
Bibliografische Information der Deutschen Nationalbibliothek: Die Deutsche Nationalbibliothek verzeichnet diese Publikation in der Deutschen Nationalbibliografie; detaillierte bibliografische Daten sind im Internet über http://dnb.d-nb.de abrufbar.
Alle in diesem Buch genannten Marken und Produktnamen unterliegen warenzeichen-, marken- oder patentrechtlichem Schutz bzw. sind Warenzeichen oder eingetragene Warenzeichen der jeweiligen Inhaber. Die Wiedergabe von Marken, Produktnamen, Gebrauchsnamen, Handelsnamen, Warenbezeichnungen u.s.w. in diesem Werk berechtigt auch ohne besondere Kennzeichnung nicht zu der Annahme, dass solche Namen im Sinne der Warenzeichen- und Markenschutzgesetzgebung als frei zu betrachten wären und daher von jedermann benutzt werden dürften.

Verlag: Südwestdeutscher Verlag für Hochschulschriften Aktiengesellschaft & Co. KG
Dudweiler Landstr. 99, 66123 Saarbrücken, Deutschland
Telefon +49 681 37 20 271-1, Telefax +49 681 37 20 271-0, Email: info@svh-verlag.de
Zugl.: Lausanne, Ecole Polytechnique Fédérale de Lausanne EPFL, Diss., 2009

Herstellung in Deutschland:
Schaltungsdienst Lange o.H.G., Berlin
Books on Demand GmbH, Norderstedt
Reha GmbH, Saarbrücken
Amazon Distribution GmbH, Leipzig
ISBN: 978-3-8381-0563-5

Imprint (only for USA, GB)
Bibliographic information published by the Deutsche Nationalbibliothek: The Deutsche Nationalbibliothek lists this publication in the Deutsche Nationalbibliografie; detailed bibliographic data are available in the Internet at http://dnb.d-nb.de.
Any brand names and product names mentioned in this book are subject to trademark, brand or patent protection and are trademarks or registered trademarks of their respective holders. The use of brand names, product names, common names, trade names, product descriptions etc. even without a particular marking in this works is in no way to be construed to mean that such names may be regarded as unrestricted in respect of trademark and brand protection legislation and could thus be used by anyone.

Publisher:
Südwestdeutscher Verlag für Hochschulschriften Aktiengesellschaft & Co. KG
Dudweiler Landstr. 99, 66123 Saarbrücken, Germany
Phone +49 681 37 20 271-1, Fax +49 681 37 20 271-0, Email: info@svh-verlag.de

Copyright © 2009 by the author and Südwestdeutscher Verlag für Hochschulschriften Aktiengesellschaft & Co. KG and licensors
All rights reserved. Saarbrücken 2009

Printed in the U.S.A.
Printed in the U.K. by (see last page)
ISBN: 978-3-8381-0563-5

To my parents

Contents

Abstract		ix
Acknowledgements		xi
Glossary		xiii

1 Introduction 1
    1.1 The Field of Machine Tools 1
        1.1.1 Today's Market Demands 3
    1.2 The Different Types of Kinematics 6
        1.2.1 The Gruebler Criterion: Mobility of a kinematics in space 7
        1.2.2 Serial Kinematics 9
        1.2.3 Parallel Kinematics 10
        1.2.4 Hybrid Kinematics 11
        1.2.5 The left/right hand concept 13
    1.3 Summary of the Chapter 14

2 Aims of the Project 15
    2.1 General Objectives 15
    2.2 Contributions and Originalities 15
    2.3 Postulate 16
    2.4 Environment and Limits of the Thesis 16
    2.5 Organization of the thesis 17

3 Analysis of the Advantages and Limits of Parallel Mechanisms ... 19
    3.1 Advantages 19
        3.1.1 Stiffness 19
        3.1.2 Mobile Mass 21
        3.1.3 Dynamics 21

CONTENTS

	3.1.4	Eigenfrequencies	22
3.2	Limits		25
	3.2.1	Limited Angular Amplitudes	25
	3.2.2	Mechanism Complexity	27
	3.2.3	Non-intuitive Movements	28
	3.2.4	Complex Mathematical Modeling	29
3.3	Summary and Conclusion of the Chapter		30

4 State of the Art: Hybrid Kinematics for Machining Tasks   **31**
- 4.1 Kinematics 31
  - 4.1.1 Degree of Parallelism = 2 35
  - 4.1.2 Degree of Parallelism = 3 39
  - 4.1.3 Degree of Parallelism = 4 46
- 4.2 Elements 47
  - 4.2.1 Interface-based Joints 48
  - 4.2.2 Flexure-based Joints 50
  - 4.2.3 Mixed Technology Joints 53
  - 4.2.4 Drive systems 54
- 4.3 Summary and Conclusion of the Chapter 56

5 Hybrid Kinematics: A Trade-Off between different Characteristics   **57**
- 5.1 Reducing the Complexity of the Kinematics 59
- 5.2 Enhancing Movement Amplitudes 61
- 5.3 Conclusion of the Chapter 62

6 Machine Elements   **65**
- 6.1 Kinematic Chains 65
- 6.2 Joints 67
  - 6.2.1 1st Evolution Spherical Joint 68
  - 6.2.2 2nd Evolution Spherical Joint 76
  - 6.2.3 3rd Evolution Spherical/Universal Joint 82
  - 6.2.4 Conclusion of the section 85
- 6.3 Segments/Rigid Bodies 86
- 6.4 Summary and Conclusion of the Chapter 90

7 Aspects of Industrialization   **91**
- 7.1 Distribution of the DOF on the kinematics 91
  - 7.1.1 Decoupling the DOF 92

	7.1.2	Identic modules	92
	7.1.3	Different, special axis characteristics	92
	7.1.4	Distribution possibilities for 3-6 axes mechanisms	93
7.2	Geometric Modeling		97
	7.2.1	Creating the Equation system	98
	7.2.2	Iterative Algorithm	100
7.3	Optimizing Kinematics		102
7.4	Protection		105
7.5	Cabling		110
7.6	Characterization Measurements		111
	7.6.1	Repeatability	111
	7.6.2	Stiffness	114
	7.6.3	Eigenfrequency	115
7.7	Summary and Conclusion of the Chapter		116

8 Design Methodology   119

8.1	Global View		119
8.2	Stepwise Description		121
	8.2.1	Application specifications	121
	8.2.2	Distributing mobilities	121
	8.2.3	Choosing kinematics	122
	8.2.4	Parameterization and optimization	123
	8.2.5	Choosing technology	123
	8.2.6	Verification	124
	8.2.7	Final design considerations	124
	8.2.8	Detailed machine design	124
	8.2.9	Assembled, controlled machine	125
8.3	Summary and Conclusion of the Chapter		126

9 Case Studies and Prototypes   127

9.1	Stewart3: Part of a hybrid 5-axes Machine for wire-EDM		127
	9.1.1	Introduction	127
	9.1.2	Project Requirements	129
	9.1.3	Distribution of Mobilities	129
	9.1.4	Kinematics	129
	9.1.5	Optimization	130
	9.1.6	Technology	131
	9.1.7	Results and Conclusion	134

CONTENTS

 9.2 MinAngle: A hybrid 5-axes High Precision Machine Tool for μEDM . . 135
  9.2.1 Introduction 135
  9.2.2 Project Requirements 135
  9.2.3 Distribution of Mobilities 135
  9.2.4 Kinematics 137
  9.2.5 Optimization 140
  9.2.6 Technology 145
  9.2.7 Final Design Considerations 145
  9.2.8 Results and Conclusion 146
 9.3 Omikron5: A hybrid 5-axes Machine Tool 149
  9.3.1 Introduction 149
  9.3.2 Project Requirements 149
  9.3.3 Distribution of Mobilities 151
  9.3.4 Kinematics 152
  9.3.5 Optimization 153
  9.3.6 Technology 155
  9.3.7 Final Design Considerations 157
  9.3.8 Results and Conclusion 158

10 Conclusion   **161**
 10.1 Follow-up of the Results 161
 10.2 Contributions 162
  10.2.1 Joint Family 162
  10.2.2 New Kinematics, Mechanisms and Prototypes 162
  10.2.3 Design Methodology 163
 10.3 Limits and Perspectives 163
 10.4 Final Note 164

References   **167**

List of Figures   **173**

List of Tables   **179**

Appendix   **180**

A Optimization of the MinAngle's left hand   **181**
 A.1 Crossed bars, four-bar linkage 181
 A.2 Base and Top pivots 183

A.3 Reducing the parasitic displacement	185
B Stiffness of spherical joints $\varnothing = 6, 8, 10, 12mm$	**187**
C Kinematic Models	**189**
C.1 Stewart3	189
C.2 MinAngle	192
C.3 Omikron5	196
D Catalogue of Kinematics	**201**
D.1 New Kinematics	202
D.1.1 2-axes	202
D.1.2 3-axes	203
D.1.3 4-axes	205
D.1.4 5-axes	206
D.2 Additional, Existing Kinematics	207

CONTENTS

Abstract

The machine tool industry is a well established, old and extremely important branch of today's manufacturing industry. With the ongoing globalization and the resulting increase of competition in this industry, the manufacturers have to push their technology to the limits in order to stay competitive. The architecture (*kinematics*) of most machine tools is based on a serial arrangement of joints and segments, like a human arm. The requirements regarding dynamics, stiffness and precision of these machines brought the scientists and industries to evaluate parallel kinematics for this type of application. Parallel kinematics possess a much higher potential to fulfill these demands, and they would therefore allow the access to a next level of machine performance.

Whereas the success of parallel kinematics in domains like packaging is incontestable, it proved to be less evident in machine tools. The low rotation amplitudes and the complexity of the mechanism, the main weak points of parallel kinematics, slow down the development and integration of this kind of machines.

In the last few years however, we could observe an increase in development, and more important, in the sales (1)(37)(54) of **hybrid kinematic machines**. Hybrid kinematics can, by appropriate combination of parallel and serial axes, present a well performing compromise, especially in the machine tool domain where 5 axes/mobilities and high rotation amplitudes are common.

The present document is concerned with the mechanical, industrialized design of hybrid-kinematic machine tools and their mechanical elements, and will show that

> "**Hybrid-kinematic mechanisms can outperform fully-parallel mechanisms considering all attributes for a successful and industrialized machine design.**"

The work will point out the limits of fully-parallel mechanisms and justify the use of hybrid solutions. The most important elements of the mechanisms, thereof particulary the spherical and universal joints, will be treated in a detailed manner. Industrialization

ABSTRACT

aspects will be analyzed, the difficulty for their integration will be shown, and solutions provided in order to increase the accessibility of hybrid and parallel mechanisms. A design methodology will be synthesized from all these elements and applied to three case studies. The methodology will point out important and often neglected steps and provide elements and tools to support the designer in the whole process of creation.

Furthermore, by providing a broad catalogue of both new and existing hybrid and parallel kinematics, this work is intended to stimulate and inspire the creativity of the designer.

The three final cases studies, each differing in their application domain and representing each an unpublished concept, will illustrate and validate the methodology.

The work took place around multiple industrial projects and therefore always keeps in mind the practical feasibility, with respect to an industrial environment, and the economic aspects and risks.

Keywords: *Hybrid, parallel, kinematics, machine tool, precision, stiffness, dynamics, industrialization, spherical joints, universal joints, catalogue, EDM, deburring.*

Acknowledgements

First, I would like to express my gratitude to my thesis director Prof. Reymond Clavel. With his profound knowledge about robotics and his ingenious ideas he contributed a lot to this work. His incessant disposition to discuss problems on one hand, his trust and the resulting autonomy on the other hand, set optimal working conditions for this thesis.

All the projects lead during this thesis were the result of profitable collaborations as a team. For this reason I would like to thank my direct colleagues, Yves-Julien Regamey, Maurice Fracheboud and Bérangère Le Gall for their contributions and the enjoyable collaboration.

Additionally, several excellent students accomplished their master thesis or semester projects on my responsibility, working on the industrial projects related to this thesis. They all made a contribution to the success of these projects. For these reasons I would like to thank especially Yves Allemand, Jürg Marti, Philipp Kobel and Vincent Fahrny. Furthermore, I'd like to thank Willy Maeder for his countless advices concerning mechanical design issues, Mohamed Bouri for the rapid implementations of his own, well-performing control solutions, and Lionel Beerens for his fast and effective IT support. All three are part of the backbone of the laboratory, and they support each project with their valuable competencies.

The different prototypes wouldn't have been financed nor manufactured without the motivation and engagement of all industrial partners. Therefore, i would like to thank all partners I collaborated with, especially:

- Tomislav Matievic of Mikron SA Agno, for his help, enthusiasm and initiative leadership.

- Andrea Marcoli of Mikron SA Agno, for his valuable contributions to the mechanical design of the Omikron5.

- Ivano Beltrami and Karl Tobler of GF AgieCharmilles SA, for their support in developing the Stewart3.

ACKNOWLEDGEMENTS

- Stefano Bottinelli and Yann Mabillard of Mecartex SA, for their help and manufacturing of the MinAngle.

I also would like to express my gratitude to the thesis jury, Dr Ivano Beltrami, Dr François Pierrot, Prof. Paul Xirouchakis and Prof. Juergen Brugger, for their attentive lecture of the work and their constructive comments. Furthermore, a very special thank to Beat Brühwiler and Christian Schneider. They did a immense job correcting all the manuscript and making suggestions for improvements.

I would like to take the opportunity to thank all my friends which supported and encouraged me during this work. Thanks a lot especially to Dominique for the recreative time spent sailing on the lakes and the constructive discussions if the wind was failing. A special thank to my beloved Paola, for her tolerance and hearty support during the tough times at the end of the PhD.

And finally I would like to thank my parents. They always trusted me, gave me the chance to do whatever project I had in mind, and supported me through it. For these reasons I am very grateful and dedicate this work to them.

Glossary

Active joint An active joint is actuated by any kind of motor. The typical serial robot possesses only active joints. See also *passive joint*.

Chip removal rate The percentage of material removed from a workpiece with respect to the initial mass.

Chip-to-chip time The time consumed between two machining tasks due to a tool changing.

Compliance The inverse of stiffness.

Degree of freedom (DOF) Any non-constrained object in space possesses 6 degrees of freedom, 3 translations along the cartesian axes X, Y and Z, and 3 rotations around the same axes $(\theta_x, \theta_y$ and $\theta_z)$. They form a set of independent variables that are used to describe the spatial position and orientation of an object.

Degree of parallelism This notion, only applicable to hybrid kinematics, determines the amount of DOF (m) which are performed by the parallel kinematics of a hybrid machine (with n total mobilities, therefore $m < n$).

Dexterity Concerning machines, the qualitative term dexterity stands for the capacity/amount of movement, therefore referring to the amount of mobilities and the movement ranges.

End-effector The end-effector, or output, of a robot is the end of the kinematics where the application-specific tool is mounted.

Footprint Normally known as the 2-dimensional surface that a certain object requires, it is more generally used to describe the volume that a certain object requires.

Geometric parameter A variable that represents a relevant geometric quantity in a mechanism (e.g. length of a strut, position coordinate of a joint).

Kinematic chain A kinematic chain is an assembly of joints and their connecting rigid segments. The assembly is linear and forms a chain connecting the input to the output. A serial kinematics consists only of 1 kinematic chain, a parallel kinematics consists of multiple chains.

Kinematic Scheme A special scheme used to illustrate the kinematics of a mechanism. It shows the order and type of joints and structural elements, but it does not give any information about the orientation and geometry of the elements.

Kinematics The kinematics (plur.) define the type, quantity and the relative placement of all the elements (joints, structural elements) of a mechanism. It is also often called the *architecture* or *topology*.

Mechanism A system of interacting bodies and joints which is designed to perform the transmission of forces and movements.

Mobility A mobility is a Degree of Freedom (DOF).

Passive joint A passive joint is non-actuated. Its movement is coupled to the other joints.

Pose, posture, position Describe, with respect to all 6 DOF, the state of an object in space. These terms are often used in relation to the end-effector of the robot.

GLOSSARY

RCM, Remote Center of Movement A rotation takes place around an offset center. Physically, no joint is located in this center.

Robot A robot is a mechanical or virtual, artificial agent. It is usually an electromechanical system, which, by its appearance or movements, conveys a sense that it has intent or agency of its own. In an industrial context a robot is an actuated automatic multi-axes mechanism.

Robotics Robotics is the science and technology of robots, their design, manufacture, and application.

Singularity Singularties occur in certain postures (called singular configurations) of the mechanism and either limit the mechanism's motion capabilities or allow an unwanted and uncontrolled movement of the mechanism (74).

TCP Tool Center Point. Used to designate the outer extreme of a machine, at its output.

Workspace The volume which can be covered by the tool. The workspace, for mechanisms performing rotations, also includes the angular strokes.

1
Introduction

1.1 The Field of Machine Tools

The machine tool is often called *the mother of machines* because of its place in our production industry (73). It's our industry's tool to produce parts of other machines, vehicles, electronic elements, tools and so much other objects. In the life cycle of many products it stands at the beginning, either fabricating the product itself or just a part of it, or even just machining components of the future production line. It can be just the tool to create other tools, but at the end, it's **the essential core of every production system**.

A machine tool *adds value and/or functionalities to an initial, brute workpiece by modifying its shape, creating by this the final, refined workpiece.*

The term *machine tool* consists of the two words *machine* and *tool*, thereby differing from the manual crafting. It is a tool that is guided by a mechanism, the *machine*, and its goal is to process or manufacture an objet. As soon as this guided mechanism is also automatically driven (by any kind of artificial control) we are in the domain of robotics. Therefore, a welding robot and a CNC-mill are both considered as machine tools in this work, and there is no intended difference between a machine tool and an industrial robot.

The domain of machine tools, because of its broad definition, includes a lot of different processes which can be subdivided in several categories. Table 1.1 gives an non-exhaustive list of processes.

1. INTRODUCTION

Table 1.1: A non exhaustive list of processes executed by machine tools.

Category	Process
Machining	Turning, milling, drilling, deburring, grinding, polishing, engraving, water-jet cutting
Optical	Laser cutting, laser welding, combined laser- and water-jet cutting
Electro-chemical	Electro-discharge machining (EDM)
Opto-chemical	Stereo-lithography, laser sintering
Forming	Stamping, forming, folding

Figure 1.1 illustrates two examples of different machine tools.

(a) Laser cutting machine by TRUMPF (www.trumpf.com)

(b) Milling machine during a surfacing process (picture by www.epma.com)

Figure 1.1: Examples of machine tools

The machine tool is a very complex system which includes several different domains reaching from mechanics, mathematics, physics, metallurgy, thermal behavior and numerical control. Numerical control itself includes electronics, control and informatics. Then, there is also the machining process itself – taking place between machine and workpiece – which can enclose different aspects. All these domains cohabit in machine tools, making it a very vast and interconnected domain of research and development.

1.1 The Field of Machine Tools

The placement of the machine tools in the production chains implicates a very high *quality and reliability* in their processes. As they are production tools, and intended to create parts of other machines or mechanisms, they need to guarantee very *high precisions and a long durability*. These very high demands, of course, induce an important price. Therefore, a machine tool is often a fundamental investment for a company.

Not seldom it can be heard that "the machine tool manufacturer is his own major competitor" because of their machines lasting for years keeping still very good results, and this, of course, does not cause the client to buy a new machine. Additionally to this, the more and more increasing productivity of the newer machines enables the buyer to subsist with a constant quantity of machines, even if his production slightly increases. All these elements lead to the fact that the machine tool industry is a very tough and slowly developing domain. The machine tool manufacturer will in general react conservatively about the intention of integrating the most modern technologies in their machines.

1.1.1 Today's Market Demands

The workpieces manufactured on machine tools are undergoing significant changes in the last few years. This chapter will explain these changes and explain the new needs in the domain of machining. In order to do this, we will take a global view on today's market demands for any products.

Increasing competition and globalization challenges the industry to manufacture new products within a shorter time to market, in a bigger variety and in a more efficient manner than some years ago. Once on the market, the products of today expect a shorter life time and are soon replaced by newer versions with superior capabilities. This considerably reduces the typical batch sizes and makes high demands on the *flexibility of the machine tool* in order to rapidly adapt to the next workpiece.

Another point is the *miniaturization and the augmenting demands in precision*. The pieces tend to get smaller and the resulting needs in precision get higher.

Since the number of elements/parts of a product is highly defining its costs, today's products consist of a minimum of elements. Those elements have to include *more functionality and naturally grow more complex* (54). Large structural parts of modern aircraft for example present chip removal rates (37) of more than 95% because of their

1. INTRODUCTION

complex shapes. Today, a lot of such parts are manufactured on hybrid machine tools (21)(37). The evolution tends towards the *complete integrated manufacturing* of these more and more complex parts. By integrated manufacturing we are thinking of the complete ensemble of manufacturing steps necessary to obtain the finalized workpiece.

Figure 1.2: Examples of multiple turbine blades machined as a single part using an ECM process (www.barber-nichols.com)

The manufacturing tools need to evolve as well if the industry wants to follow this trend. The tools need to be faster, more flexible, more modular and allow the manufacturing of more complex parts, in other words, the machines need more capabilities and dexterity. Serial machine tools are getting to their limits (regarding dynamics and stiffness) and there is a need for more innovative solutions in the domain of parallel and hybrid mechanisms. This is a developing field in the domain of machine tools. The theoretical advantages of using this technology are obvious, nevertheless, this development is happening very slowly because of certain unresolved problems and the prudent attitude of the industry.

It is all the more important to follow this trend in the machine tools industries of the western countries, where some of the former market leaders (Great Britain, France, USA) are considered today as underdeveloped[1] in this sector (63). They need to develop

---
[1] With respect to other industries.

1.1 The Field of Machine Tools

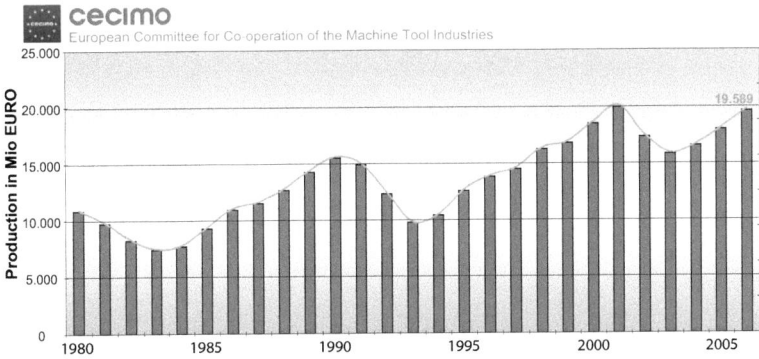

Figure 1.3: Market survey, CECIMO Countries: Yearly production 1980-2006. (Source: CECIMO (3)). Cecimo countries include the 15 most important producers in Europe.

market niche technologies with high added technological value and/or high productivity potential if they want to compete with emerging low-salary countries (China, south-east Asia in general). Concerning Switzerland, it is since several years by far the leader[1] (per resident) in manufacturing, exporting, importing and consuming machine tools (63).

Figure 1.3 illustrates the yearly production of machine tools in Europe. The development of this market follows the general economic growth and does not illustrate any striking peaks, which is a sign for an established market.

Figure 1.4 shows the global repartition of the machine-tool production.

---

[1] Considering the total volume of machines, Japan (followed by Germany) is the biggest manufacturing and exporting country (status 2005).

1. INTRODUCTION

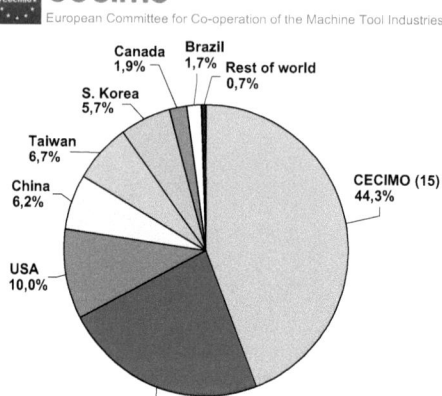

Figure 1.4: World market survey: Global repartition of production 2005. (Source: CECIMO (3). Cecimo countries represent 44% of the world-wide machine tool production.)

1.2 The Different Types of Kinematics

As already mentioned above, there exist different types of kinematics. The kinematics define the type, the relative placement and the quantity of the joints. A kinematics describes only the architecture of a mechanism, it does not give any idea of the size nor the mechanical design of this mechanism. The joints are connected to each other with rigid segments. Figure 1.5 shows the commonly-used and essential joints in robotics. We could imagine many other joints which are not mentioned in the figure but these are rarely or never used, or, can be obtained by the combination of the illustrated joints.

To illustrate a kinematics we will use a special type of figure: *the kinematic scheme*. This scheme simplifies a mechanism to its basic components, the articulations and the rigid segments (which interconnect the articulations). The kinematic scheme does not include information about actuation or geometric dimensions. The application-specific tools will be illustrated by simple grippers. For instance, in a kinematic scheme, there is no visible difference between an EDM tool and a milling spindle. A kinematic scheme of a serial-kinematic robot is illustrated in figure 1.6.
In the case where some more spatial information is needed we will expand these schemes,

1.2 The Different Types of Kinematics

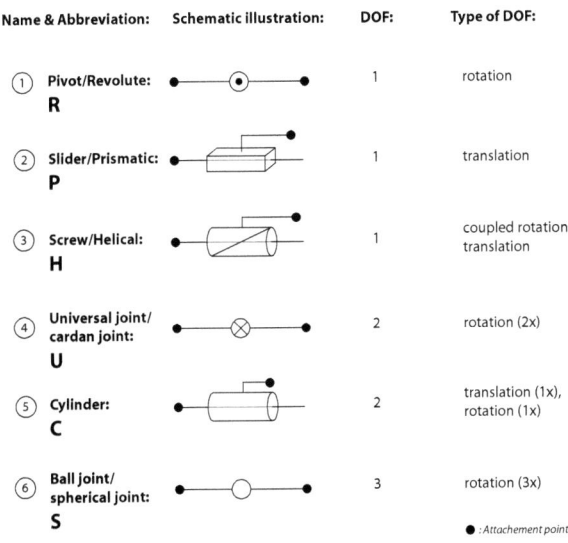

Figure 1.5: A definition of the most-used basic joints. Kinematic chains can be defined by the sequence of their joints, e.g. UPS (universal-prismatic-spherical).

the spatial relative placement and the orientation of the joints will be added.

1.2.1 The Gruebler Criterion: Mobility of a kinematics in space

Gruebler's formula (33) determines the amount of *degrees of freedom or mobilities*[1] of a kinematics in space.

A non-constrained part in space possesses 6 mobilities. Therefore, the amount of mobilities of a kinematics can be defined by:

$$Mo = 6n - \sum_{i=1}^{k} GF_i \qquad (1.1)$$

Where n is the number of parts of the kinematics, k the number of articulations and GF_i is the number of *generalized forces or constraints*[2] for the articulation i. The

---
[1] These equal terms will both be used in this work.
[2] Both terms will be used.

1. INTRODUCTION

number of generalized forces is the complement to 6 of the joint's mobilities, therefore:

$$Mo_i = 6 - GF_i \qquad (1.2)$$

where Mo_i are the mobilities of the articulation i. These values, with respect to a joint type, can be consulted in figure 1.5.

Using equations 1.1 and 1.2 we obtain:

$$Mo = 6(n - k - 1) + \sum_{i=1}^{k} Mo_i \qquad (1.3)$$

where the "-1" in $6(n-k-1)$ comes from the fact that the base of the kinematics is considered as a fixed part in space. If we don't consider the base of the kinematics and only count n_m as the number of mobile parts the equation can be simplified to:

$$Mo = 6(n_m - k) + \sum_{i=1}^{k} Mo_i \qquad (1.4)$$

The Gruebler criterion is a very useful tool to calculate rapidly the mobilities of any kinematics and verify if it's isostatic or hyperstatic/overconstrained:

Isostatic The calculated mobilities of the robot's output is equal to the intended number of mobilities at the output.

Hyperstatic The calculated mobilities of the robot's output is lower than the intended number of mobilities at the output, the kinematics is therefore often designated as **overconstrained**. This occurs because of a conflict in the definition and amount of the constraints. A kinematics that is realized to accomplish p mobilities, but, according to the Gruebler criterion only possesses q mobilities (where $q < p$) is called $(p - q)$-times **overconstrained**.

In order to demonstrate the use of this criterion we will apply it in section 1.2.3 on the well-known Delta robot.

1.2 The Different Types of Kinematics

1.2.2 Serial Kinematics

A serial kinematics is formed by one single chain of elements that connects the base to the end-effector of the robot, and which does not form any loop, therefore often called *open-loop kinematics*. All elements that compose the kinematics are placed one after the other in series, comparable to a human arm.
They are the most common and oldest type of kinematics used today, most machines are build after this scheme.

The figure 1.6 shows a robot with a typical serial kinematics. Its only kinematic chain is easily visible in the picture and in the kinematic scheme.

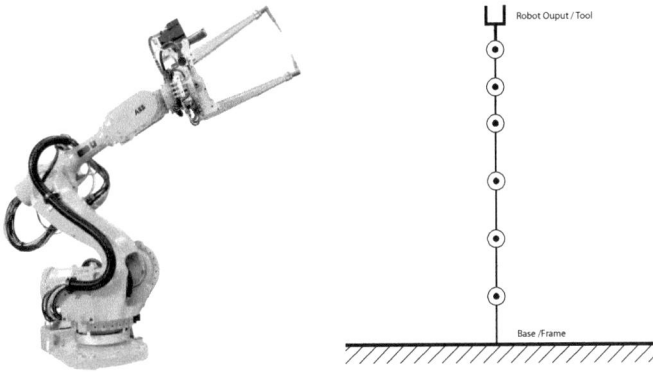

(a) A serial-kinematic robot from ABB.

(b) Kinematic scheme of the ABB IRB 6600ID. The robot possesses 6 DOF.

Figure 1.6: A serial-kinematic robot. The ABB IRB 6600ID (www.abb.ch), used for welding in the automotive industry.

For serial kinematics, the sum of the mobilities of all joints is equal to the mobilities of the mechanism[1]. The robot in figure 1.6 has 6 joints, each having 1 mobility, and this is equal to the number of mobilities of the whole mechanism.

---

[1] Some few exceptions exist, e.g. when using four-bar-mechanisms to maintain parallelism of certain segments.

1. INTRODUCTION

1.2.3 Parallel Kinematics

A parallel kinematics, as opposed to a serial kinematics, is *formed of multiple kinematic chains* that implicitly create loops. Thus, these types of architectures are often called *closed-loop kinematics*. In contrast, serial kinematics are called open-loop kinematics.

The figure 1.7 shows one of the most known parallel kinematics.

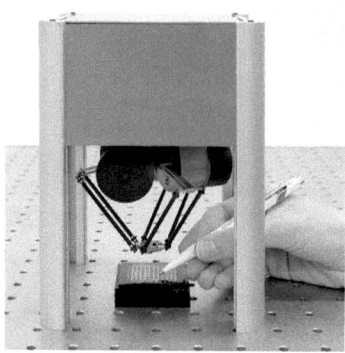
(a) Picture of the CSEM PocketDelta.

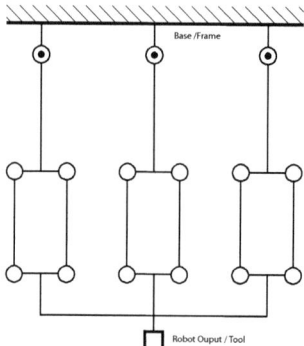
(b) Kinematic scheme of the CSEM Pocket-Delta.

Figure 1.7: A parallel-kinematic robot. The PocketDelta robot from the CSEM (www.csem.ch), designed for the integration in a micro-factory. This robot is based on the well-known Delta kinematics developed by Clavel (15) and patented in 1986 (14).

By applying the Gruebler criterion (equation 1.4) on this kinematics we obtain:

- $n_m = 10$ mobile parts

- $k = 15$ articulations

- $\sum_{i=1}^{k} Mo_i = 3*1 + 12*3 = 39$ the sum of the individual mobilities, 3 pivots with 1 mobility, and 12 spherical joints with 3 mobilities each.

$$Mo = 6(n_m - k) + \sum_{i=1}^{k} Mo_i = 6(10-15) + 39 = 9 \qquad (1.5)$$

The result is 9 mobilities in total. This number does not represent the DOF of the output, it does indicate the total mobilities of the whole mechanism.

1.2 The Different Types of Kinematics

In fact, the 6 bars that are each located between two spherical joints can rotate freely around their symmetry axis without changing the posture of the robot. These are so-called **internal DOF/mobilities**, they do not influence the spatial position or orientation of the end-effector.

$$Mo_{\text{end-effector}} = Mo - Mo_{\text{internal}} \qquad (1.6)$$

If we substract these internal DOF from the total mobilities we obtain the effective DOF of the end-effector, the achievable movements of the robot's tool. In this case 3.

As seen in the calculation above: **For parallel kinematics, the sum of the mobilities of all individual joints is higher than the mobilities of the mechanism.** All joints of a Delta kinematics possess together 39 mobilities, whereas the final mobilities of the mechanism are 3. This is different than for serial kinematics.

For the actuation of a parallel kinematics any of those mobilities could be driven by a motor. Normally, the designer chooses to actuate the mobilities that are the closest to the base/frame of the machine, which allows the fixation of the motor on the frame. Of course, the number of driven mobilities should correspond to the number of mobilities of the mechanism[1]. In the case of the Delta robot shown on figure 1.7, the driven mobilities are the 3 pivots that are attached to the ground.

1.2.4 Hybrid Kinematics

A hybrid kinematics is a combination of parallel and serial kinematics[2], it's neither fully-parallel nor fully serial. It consists of both open-loop and closed-loop chains.

The figure 1.8 shows one of the most known hybrid kinematics, the Tricept. Its kinematics are composed of a parallel part (the translator part or *carrier*, creating movement in space) and a serial part (the wrist, orienting the tool), the closed-loop and open-loop parts are clearly visible on the kinematic scheme.

---

[1] The robot is identified as being redundant when the amount of driven mobilities is higher than the mobilities of the mechanism. In the opposite case the robot is called underactuated or undetermined.

[2] The serial arrangement of two parallel modules constitutes a hybrid kinematics.

1. INTRODUCTION

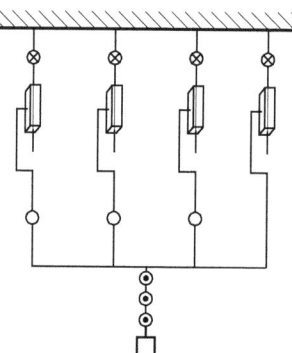

(a) Picture of the ABB IRB 940 Tricept.

(b) Kinematic scheme of the ABB IRB 940 Tricept

Figure 1.8: A 6-axes hybrid-kinematic robot from ABB, used for machining tasks. The ABB IRB 940 Tricept (www.abb.ch). This robot is based on the well-known Tricept kinematics developed by Neumann (55).

1.2 The Different Types of Kinematics

1.2.5 The left/right hand concept

This chapter will introduce the left/right hand concept of a machine. This concept consists in distributing the necessary mobilities, used for a certain application, on two parts of the machine, the left and the right hand. The following definition is based on figure 1.9.

The right hand is the part of the machine going from the base to the application-specific tool

The left hand: is the part of the machine going from the base to the workpiece

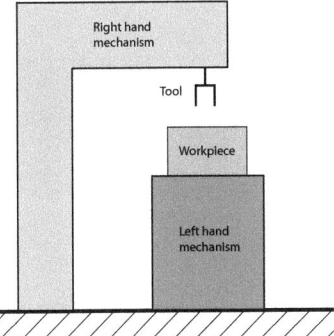

Figure 1.9: The left/right hand concept

A lot of applications in the domain of machine tools need 5 DOF, 3 translations and 2 rotations, in order to be executed correctly. These DOF can be distributed freely on both hands of the machine (since it is the relative movement between the workpiece and the tool that matters). Figure 1.10 shows two different examples of how this can be achieved.

The left hand is considered as being mounted in series to the right hand. Therefore, a mechanism that distributes the DOF on both hands can never be considered as parallel, it can be either **serial** if both hands are serial kinematics, or **hybrid**, if any hand contains a parallel part.

Later on, we will explain the importance of a correct choice concerning the distribution of the DOF.

1. INTRODUCTION

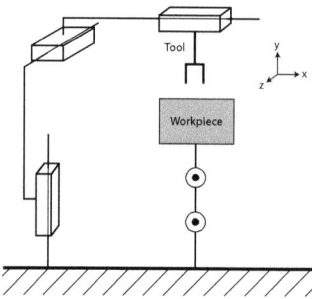
(a) Distribution of 3 translational DOF on the right hand and 2 rotative DOF on the left hand.

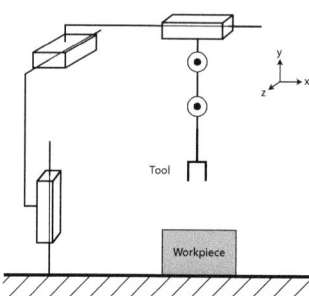
(b) All 5 DOF placed on the right hand.

Figure 1.10: Example of 2 different distributions of the DOF. In both cases the resulting DOF between tool and workpiece are 5, 3 translations and 2 rotations. Both examples are serial kinematics.

1.3 Summary of the Chapter

The chapter introduced the domain of machine tools and its importance in the production industry. A very general analysis of the evolution of products, workpieces or parts that are manufactured on machine tools was made. The different types of kinematics were introduced as well as the concept of distributing the mobilities on the 2 hands of a mechanism.

The reasons for our interest in the different types of kinematics are their respective advantages. They will be treated, as well as their respective limits, in chapter 3.

2
Aims of the Project

2.1 General Objectives

The general objectives of the present work are:

- To *show the limits of parallel mechanisms* and *justify the use of hybrid kinematics* for applications in the domain of machine tools.

- Propose a document which is *useful and inspiring* for machine designers, developers, or scientists who wish to create efficient and adapted hybrid mechanisms for certain application specifications.

- *Promotes the industrialization and accessibility* of parallel/hybrid mechanisms.

2.2 Contributions and Originalities

In order to accomplish the above-mentioned objectives, several contributions will be necessary:

- A complete *state of the art* (chapter 4) and *analysis* (chapters 3 and 5) of industrialized hybrid kinematics. The analysis will be carried out using a new performance index, the mobility inefficiency. The conclusions will show the superior performance of hybrid kinematics compared to serial or parallel mechanisms.

- Pointing out the *most influential steps*, and their links, in the machine design and synthesize them in a *design methodology* (chapter 8).

- A *catalogue of new kinematics* which helps expanding, in combination with existing solutions, the pool of kinematics (chapter 4, chapter 9 and the catalogue of kinematics D).

2. AIMS OF THE PROJECT

- An effective and well-performing proposition for very specific and determining elements of hybrid mechanisms: *The ball- and universal-joint* (chapter 6). A single joint design which allows both working modes will be proposed and its good performance will be validated through measurements. Furthermore, the modeling of the joints will be carried out, allowing the designer to adapt his design to specific requirements. A new production technique, which allows a significant increase of the joint's stiffness, will be presented.

- Solutions and propositions for all *major problems of industrialization* (chapter 7).

- Presenting new hybrid-kinematic machines in order to inspire the designer and let him take benefit from the experiences made: *3 case studies* (chapter 9), all presenting innovative kinematics and application-specific performances, will be presented. Their development through all major design steps will be outlined.

2.3 Postulate

"**Hybrid-kinematic mechanisms can outperform fully-parallel mechanisms considering all necessary attributes for a successful and industrialized machine design.**"

2.4 Environment and Limits of the Thesis

This thesis has taken place in parallel with 4 industrial projects and 4 industrial partners, resulting in 4 functional machines. All these projects were collaborations of the LSRO (Laboratoire de Systèmes Robotiques) at the EPFL (Ecole Polytechnique Fédérale de Lausanne) and Swiss companies, supported and supervised by the KTI/CTI, a federal institution for the promotion of innovation.

This work focuses on the *mechanical part of the design process*: The kinematics, the mechanical design of elements and their modeling, the optimization of structures and elements, the simulations of mechanisms.

Independently from the manufacturing process of the developed machines, this work provides solutions, always trying to enhance technical and economical aspects of a machine. These aspects are summarized in figure 2.1.

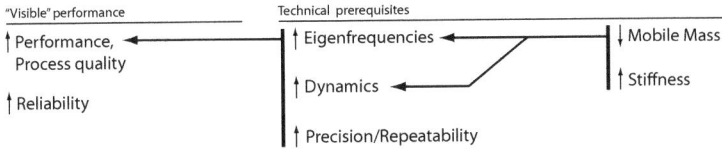

Figure 2.1: Targeted technological and economic aspects as well as their links. The small vertical arrows describe the general, desirable effect on the respective aspect or characteristic of a machine tool. The economical and technical performance indicators (the "visible" performance of a machine) are listed on the left. The technical prerequisites are listed on the right. The association between the different aspects is represented by horizontal arrows.

2.5 Organization of the thesis

Chapter 1 introduces most of the notions used in this present work. A short market survey about today's tendency in the machine tool sector is presented. The different types of kinematics are presented.

Chapter 2 presents the aims of the project and their premises, the originalities, the postulate and the scientific contributions. It also presents the conditions in which the project took place.

Chapter 3 enumerates the advantages and disadvantages of parallel kinematics, with respect to serial kinematics. The factors leading thereto are accentuated.

Chapter 4 presents a state of the art of published hybrid-kinematic machines, synthesis methods and machine elements.

Chapter 5 shows the advantages and possibilities obtained when using hybrid-kinematic machines. It presents how hybrid kinematics can deal with the problems of fully-parallel kinematics and shows their efficiency when it comes to kinematics with high dexterity.

Chapter 6 compiles a catalogue of machine elements for hybrid mechanism. It presents a completely new concept for high-performance ball- and universal-joints as well as a simple and effective method to increase a joint's stiffness. Furthermore, the modeling of the joints is carried out and validated with measurements.

2. AIMS OF THE PROJECT

Chapter 7 emphasizes the aspects of industrialization for different elements of hybrid or parallel machine tools. It shows the necessity for a *design for industrialization* of all aspects of the machine, from the kinematics to the geometric models.

Chapter 8 presents the developed design methodology for hybrid kinematics.

Chapter 9 presents 3 developed and realized machine tools that will serve as case studies to validate the design methodology. All 3 machines were developed with different specifications and for different industrial applications.

Chapter 10 concludes this work. It summarizes the results and presents future research possibilities.

3

Analysis of the Advantages and Limits of Parallel Mechanisms

3.1 Advantages

Parallel kinematics possess many advantages over the older and more established serial kinematics, nevertheless they are still lagging behind in their integration in today's industry. Their advantages will be enumerated in this section whereas the disadvantages and limits will be examined in section 3.2.

3.1.1 Stiffness

Stiffness is the resistance of an elastic structure to deflection or deformation by an applied force. It is defined as:

$$k_x = \frac{\delta F}{\delta x} \qquad (3.1)$$

where F is the applied force and x the deflection, both according to a same direction. Its standard unit is $[\frac{N}{m}]$ or $[\frac{N}{\mu m}]$.

Stiffness is a very important characteristic in machine tools and has to be increased as much as possible. It greatly influences the precision, surface finish and, as we will see later, the achievable dynamics of the machine.

Because of their architecture, the parallel kinematic machines (PKMs) offer a theoretically higher potential for stiffness compared to serial kinematics. The single structural segments of a serial kinematic machine are forced to bear bending and/or torsion modes. These modes can be limited or even avoided in certain cases of parallel kinematics.

3. ANALYSIS OF THE ADVANTAGES AND LIMITS OF PARALLEL MECHANISMS

The figure 3.1 compares two, in terms of movement capabilities equal mechanisms, one of these representing a serial kinematics and the second a fully-parallel kinematics.

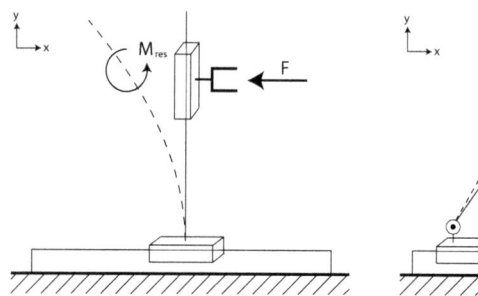 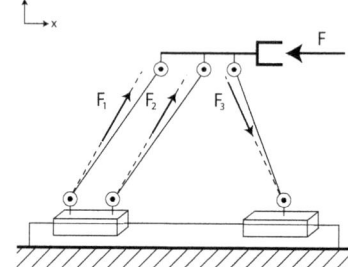

(a) Serial-kinematic mechanism loaded by a lateral force. Bending moment M is generated.

(b) Parallel-kinematic mechanism loaded by a lateral force. Traction and compression forces F_i are generated along the struts.

Figure 3.1: These two mechanisms have the same movement capabilities, both are translators in the $X - Y$ plane possessing 2 DOF.

Several facts can be pointed out:

Deformation: The deformation quantity is much higher in a bending or torsion mode that in the traction-compression mode. Parallel-kinematic mechanisms are not imperatively working only in traction-compression, but when designed optimally this can be achieved.

Shared Load: The parallel-kinematic mechanism, by definition, possesses multiple kinematic chains, and these are sharing the external load.

Number of articulations: For a same movement capability a parallel-kinematic mechanism will always have more articulations than his serial-kinematic equivalent.

As mentioned above, a parallel-kinematic mechanism will always have more articulations than its serial equivalent. Hence, the articulations need a special attention because they are very sensitive elements concerning stiffness. The section 6.2 is therefore dedicated to their design.

3.1 Advantages

3.1.2 Mobile Mass

The mobile mass are all masses moving during any operation of the mechanism. Parallel or hybrid mechanisms present the advantage of having less mobile mass than serial kinematics. As discussed in the section 1.2.3, the designer mostly tries to actuate the joints that are the closest to the base of the robot. Thanks to this, the motors and their eventual gearboxes – mostly representing the heaviest parts of a robot – can be fixed on the base and are therefore not moved during the operation.

Also, the structural elements are presenting a better ratio weight/stiffness, as they are often working in an optimal traction/compression mode.

Reducing the mobile masses of a robot brings an advantage (depending on the case):

- *Higher dynamics of the machine.* With the same power and a lower mobile mass the robot can be driven faster.

- *Less energy wastage.* Less power is used for the same acceleration of the robot. An ecological value for some people, mostly a financial reason for the industry. Furthermore, less heating is generated. This can reduce the stress and deformation induced by the heating.

3.1.3 Dynamics

Enhanced dynamics are directly resulting from the reduction of the mobile mass (the driven load). An equal motor torque will accelerate a lighter mass faster, if the transmission ratio is adapted to its new optimum. Transmission ratios decrease for lighter loads.

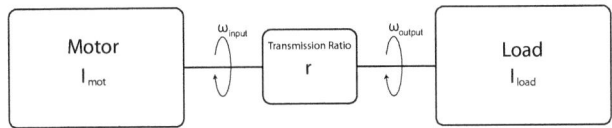

Figure 3.2: The figure shows a schematic view of a drive system. The system is composed of a motor, a gearbox with transmission ratio r and a driven load.

Let us assume that the inertia of the load is higher than the inertia of the motor. This is actually true for most serial kinematic machines. In order to achieve highest possible acceleration (given a certain motor torque and load) the transmission ratio r

3. ANALYSIS OF THE ADVANTAGES AND LIMITS OF PARALLEL MECHANISMS

needs to be chosen carefully. In fact, optimal actuation is achieved when the inertia of the load I_{load}, seen from the motor-side, is equal to the inertia of the motor I_{motor} (17). This gives us the following equation:

$$I_{motor} = I_{load}\frac{1}{r^2} \qquad (3.2)$$

Or, if defining the transmission ratio:

$$r = \frac{\omega_{input}}{\omega_{output}} = \sqrt{\frac{I_{load}}{I_{motor}}} \qquad (3.3)$$

Where I_{motor} includes all elements of the drive system on the motor side, which means in front[1] of the effective transmission r. For example a wave-generator of a Harmonic Drive or a bellow coupling, both having significant inertias which are not negligible.

As we can see from equation 3.3, if the inertia of the load is decreasing and approaching I_{motor} the transmission ratio is tending towards 1. A driving system with transmission ratio $r = 1$, and where the actuator is directly attached to the load, is called **direct drive**.

On very lightweight and dynamic parallel mechanisms we can even observe the inverse phenomenon. The mobile mass is so low (and the targeted dynamics so high) that the motor inertia I_{motor} is becoming the predominant part of the total inertia.

The designer has to take into account all elements when dimensioning the drive system of a dynamic machine, also on the motor side, since this side tends to be decisive for highly dynamical machines. In consequence, elements like couplings, ballscrews or bearing housings can have an important effect on system dynamics.

As a consequence of their high dynamics, parallel mechanisms allow to work with faster cycle times. Downtime can be reduced[2], and, if not limited elsewhere, machining speed can be increased.

3.1.4 Eigenfrequencies

The Eigenfrequencies are the natural, inevitable mechanical frequencies at which the mechanism tends to vibrate. Low Eigenfrequencies can have a very negative influence

---
[1] Left side of figure 3.2.
[2] For example when transferring between 2 machining operations or during a tool-change.

3.1 Advantages

on a machine itself, the processed workpiece and even the operator (73):

Dimensional and geometric accuracy of the workpiece and surface finish: The dimensions as well as the geometry of the workpiece can be spoiled by a vibration, even if its amplitude is very low. A surface finish that has been machined, while the robot was vibrating, will leave defects visible even by the human eye[1].

Durability of machine elements: Constant vibration risk to damage the sensitive elements of a machine. Bearings – or in general joints – have only small contact regions between elements. These regions can be damaged because of the repeatedly high contact pressures due to vibrations.

Productivity #1: Some applications like machining are limited in their productivity because of low Eigenfrequencies. A certain cutting depth cannot be exceeded because of emerging vibrations. This limits the productivity, the machines have to process several more cuts for a given cutting depth.

Productivity #2: High accelerations are needed for a fast displacement of the robot. The problem is that high accelerations can create high-frequency mechanical excitations which can lead to vibrations around the Eigenfrequency. Depending on the smoothness of the movement law[2] (constant acceleration, constant jerk etc.) an Eigenfrequency can be excited (24). A low Eigenfrequency can therefore limit the choice of movement law.

Durability of machining tools: Modern machining tools are made of ultra-hardened metals or even ceramics in certain cases. Such material are needed to guarantee a long-lasting cutting quality but they are extremely fragile and very sensitive to vibrations.

Operator: Vibration or the noise emitted by the machine may be a strain on the operator who is constantly exposed to.

Vibrations of a mechanism cannot be completely avoided because of the elasticity of their elements, but their location in the frequency domain can be modified and be placed higher. For this, the design parameters which influence this location will be derived from a simple model.

---

[1] The human eye can be very sensitive for qualifying a surface finish. Orienting the surface in a favorable angle to a light source makes the defects easily visible

[2] Which was chosen for the trajectory generation of the toolpath

3. ANALYSIS OF THE ADVANTAGES AND LIMITS OF PARALLEL MECHANISMS

Toenshoff (73) says that every single Eigenfrequency of a mechanism can be closely modeled by a simple 1-dimensional harmonic oscillator. The damping of the oscillator is not taken into account, the goal of this section being to isolate the parameters that influence the location of the 1st Eigenfrequency. Figure 3.3 shows a free harmonic oscillator that will be used to localize the parameters:

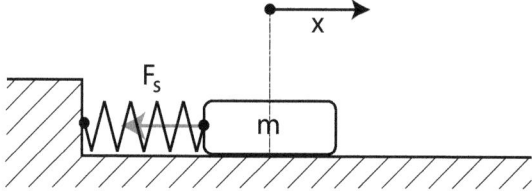

Figure 3.3: The figure shows a schematic view of a 1-dimensional free, frictionless harmonic oscillator. F_s is the force carried out by the spring, m the mobile mass.

Friction force will be neglected since it is an important concern for machine tool designers to have as little friction as possible. Applying Newton we obtain the following equation:

$$\sum F = ma = F_s = kx \qquad (3.4)$$

$$0 = kx - m\ddot{x} \qquad (3.5)$$

where k is the spring constant[1] and x is the coordinate describing the movement. Equation 3.5 is a second-order homogeneous differential equation with the solution of the type:

$$x(t) = X_m sin(\sqrt{\frac{k}{m}}t + \phi) \qquad (3.6)$$

The solution is a periodic function with a pulsation $\omega_0 = \sqrt{\frac{k}{m}}$, the oscillating frequency therefore becomes $f_0 = \frac{1}{2\pi}\sqrt{\frac{k}{m}}$.

This short modeling shows that the mass m and the stiffness k are the significant parameters which determine the placement of the first Eigenfrequency. High Eigenfrequencies can be achieved by lowering the masses and enhancing the stiffness, exactly

---
[1] the stiffness of the spring

the principal advantages of parallel/hybrid kinematics.

High accelerations excite the machine structure up to high frequencies. The 1st Eigenfrequency therefore needs to placed as high as possible to avoid vibration of the machine structure.

3.2 Limits

The precedent section showed the advantages of parallel mechanisms over the conventional serial mechanisms. In contrast, this chapter will inform about the limits of fully-parallel mechanism and will introduce how hybrid mechanisms can overcome these limitations.

3.2.1 Limited Angular Amplitudes

Most parallel kinematics present a lack of rotation amplitude compared to serial kinematics which can easily achieve complete 360° rotations. This comes from the fact that parallel kinematics, because of their constitution with multiple kinematic chains, need to work in a differential manner on multiple kinematic chains to control a torque, the effort which will create the rotation. A certain amount of rotation (on the passive joints) cannot be exceeded otherwise running into a singularity. The figure 3.4 illustrates this.

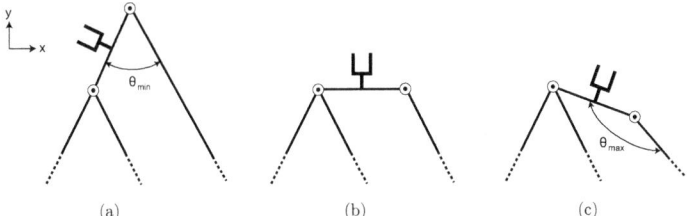

Figure 3.4: The figure shows a 3-DOF planar parallel mechanism in 3 different postures. The postures on (a) and on (c) are close to the practical limits of rotation, a few degrees further and the kinematics would reach a singularity. Totally, going from (a) to (c), the mechanism has achieved 90° of stroke. This is the typical, maximal rotation of a parallel kinematics before reaching postures where a certain level of stiffness cannot be guaranteed anymore (71).

3. ANALYSIS OF THE ADVANTAGES AND LIMITS OF PARALLEL MECHANISMS

The typical maximum rotation for parallel kinematic, due to their singularities, lies around 90°. This amplitude is limited by the angles performed on the single joints and the resulting orientation of the segments (as visible on figure 3.4). Exceeding these angles leads either to the loss of control, or to the loss of a mobility of the end-effector (74).

There exist some mechanisms (Hita STT (71), Alpha5 (60), the Orthoglide 5-axes (13)(Appendix D.2, figure D.10), the Par4 (52) (chapter 4, figure 4.15)) which can go beyond these limits by using specialized mechanisms. The studies from Krut (44), Thurneysen (71) and Nabat (52) analyzed and proposed mechanisms which can generate high-amplitude rotations on parallel kinematics. Figure 3.5 shows the Hita STT (72) machine tool that can achieve about 120° rotation by using a four-bar linkage similar to how excavator blades are actuated.

(a) An image of the Hita STT Prototype.

(b) A model of the 4 parallel axes of the Hita STT.

Figure 3.5: The Hita STT mechanism (71). One of the few parallel mechanisms that can exceed the 90° rotation amplitudes.

Concerning movement amplitudes in general – translations and rotations – it can be said that fully-parallel mechanisms possess less workspace, with respect to their footprints, than conventional serial kinematics. For serial kinematics the end-effector's workspace is the *combination* of the workspace of all axes, whereas for parallel kinematics it is the *intersection* of the workspace of the single kinematic chains. The more parallel axes there are in a mechanism, the more the workspace is constrained and limited.

3.2 Limits

There are concepts to pass over this limit, like the collinear arrangement of the sliders in the same direction (see figure 3.5, all actuated sliders of the Hita STT machine are collinear). Thanks to this arrangement the workspace could be infinitely long – in theory – in the direction of the sliders. Unfortunately, this type of elongated workspace is only useful for few applications.

3.2.2 Mechanism Complexity

As the precedent chapter showed, there exist very innovative solutions to pass over some limits of fully-parallel kinematics. Special mechanisms can enhance the angular limits but they greatly increase the complexity of the mechanisms. A very good example of this tendency is the Alpha5 (58)(59)(60), a fully parallel robot with 5 DOF – 3 translations and 2 rotations – which can achieve ±90° on both rotation axes. This robot was developed to explore the limits of fully-parallel mechanism. It is represented in figure 3.6.

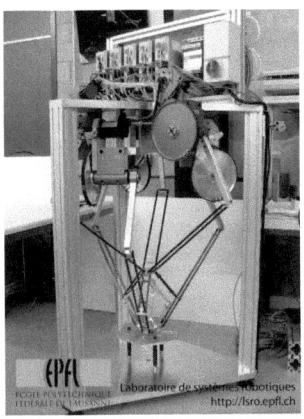 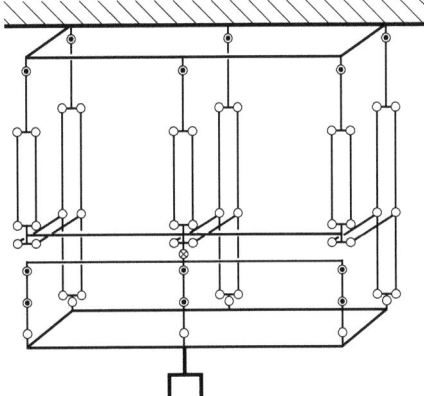

(a) An image of the Alpha5 Prototype.   (b) The kinematic scheme of the Alpha5.

Figure 3.6: The Alpha5 mechanism (59).

The sum of all mobilities $\sum Mo_i$ is equal to 140, which is a huge amount. And this, for a total of 5 DOF only. Apart from the drawback of being extremely complex, this robot can achieve high angles on 2 axes (see figure 3.7) which is extraordinary for

3. ANALYSIS OF THE ADVANTAGES AND LIMITS OF PARALLEL MECHANISMS

parallel kinematics.
The Alpha5 robot is a nice academic example of how far we can go, using fully-parallel mechanisms, but it also shows the impossibility of industrializing such a complex mechanism. The great amount of joints makes this mechanism too compliant, too heavy, and of course, too expensive.
Parallel mechanisms with 4 or more degrees of freedom tend to be very complicated in their structure, especially if they include systems to enhance their angular amplitudes.

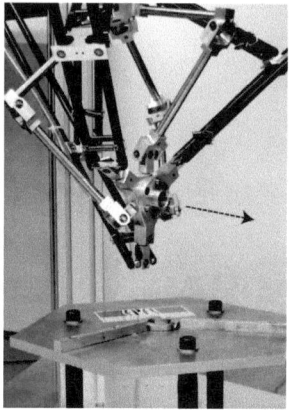

(a) An image of the Alpha5 rotated by 90° around its x axis.

(b) An image of the Alpha5 rotated by -90° around its x axis.

Figure 3.7: The Alpha5 mechanism (59) demonstrating its high angular amplitudes.

3.2.3 Non-intuitive Movements

In contrast to serial robots – especially those with cartesian axes – the end-effector's movements of a parallel kinematics cannot be assigned intuitively on the motorized axes. The simple fact of executing a straight line with the tool implies coordinated and complex movements of multiple motors. The movements of the end-effector are linked to the motorized axes through a mathematical model (see section 3.2.4) which can be very complex in certain cases.
For machine tool operators this may be confusing, as they are used to move just 1 motor and exactly predict how the end-effector will behave. Parallel kinematics need syn-

chronized and interpolated movements of all motorized axes to obtain clearly-defined, cartesian trajectories of the end-effector.

3.2.4 Complex Mathematical Modeling

For all multi-axes mechanisms the posture of the end-effector is linked to the motorized axes by more or less complex mathematic relations, the so-called kinematic models. These models are needed for a lot of different tasks going from the optimization of the mechanism to its final launch and operation.

To describe the movement of a robot we define the coordinates of the end-effector, the **output coordinates or operational coordinates**:

$$\boldsymbol{X} = \begin{pmatrix} x \\ y \\ z \\ \theta_x \\ \theta_y \\ \theta_z \end{pmatrix} \tag{3.7}$$

Where $\boldsymbol{X}$ is a vector containing the 6 independent variables to describe a posture in space. In certain cases (e.g. robots which carry out only translations) the size of the vector can be reduced to contain only the varying coordinates, the other being constant.

And the coordinates describing the positions of the actuated axes, the **articular coordinates**:

$$\boldsymbol{Q} = \begin{pmatrix} q_1 \\ q_2 \\ \cdot \\ \cdot \\ \cdot \\ q_i \end{pmatrix} \tag{3.8}$$

where i goes from $i = 1...n$, n being the number of DOF of the system. Depending on the chosen actuators, the q_i can be e.g. the angular position of a servomotor or the linear displacement of a linear actuator.

The **direct kinematic model (DK)** is a set of functions which calculates the end-effector's posture $\boldsymbol{X}$ using the articular coordinates q_i:

$$\boldsymbol{X} = f_d(q_i) \tag{3.9}$$

3. ANALYSIS OF THE ADVANTAGES AND LIMITS OF PARALLEL MECHANISMS

The **inverse kinematic model (IK)** is a set of functions which calculates the articular coordinates q_i using the end-effector's posture $\boldsymbol{X}$:

$$q_i = f_i(\boldsymbol{X}) \tag{3.10}$$

The information needed to formulate the functions is taken from the kinematics of the machines. The joints and the segments have to be transcribed in mathematic formulae using geometrical relations and trigonometry.

For parallel kinematics, these 2 models (equations 3.9 and 3.10) can be complicated[1], the functions can contain a great amount of non-linear elements. In certain cases even, the functions describing the architecture of the mechanism cannot be solved analytically. In these cases the kinematic models need to be solved numerically using an approximative algorithm that can find the roots of non-linear equations (e.g. Newton-Raphson).

The development of these models can therefore be quite complicated. The other problem that emerges from these models is the computing power needed to solve them in real-time during the operation of the robot. Even if today's computers are evolving very fast it is not always possible to implement such heavy models.

3.3 Summary and Conclusion of the Chapter

Most notions around the kinematics and characteristics of mechanisms were introduced. These notions will be used later in this document. Furthermore, the chapter presented the advantages and disadvantages of serial and parallel kinematics.

Serial and parallel kinematics have both their typical applications fields (in terms of requirements). They excel each on certain characteristics and perform less effectively on others, this, in an oppositional way. Creating hybrid kinematics by selective combination of both technologies, and making ideal use of their respective advantages, is a logic reaction.

---

[1] In general the IK is easier to obtain than the DK. In serial mechanisms however it is the opposite. Concerning hybrid kinematics, as they contain parallel- and serial-kinematic parts, they always contain the worse case for one of the parts.

4

State of the Art: Hybrid Kinematics for Machining Tasks

The state of the art will be split in two main sections, first about the kinematics, and second about the mechanical elements of the machine.

4.1 Kinematics

This section provides a short overview on the topological synthesis, a research domain that focuses on the methodological creation of kinematics. Afterwards, a complete collection of the most representative hybrid kinematics developed to date is provided.

The topological synthesis *is the method-based synthesis/creation of kinematics considering certain given specifications.*

The synthesis of kinematics has always been an emphasis of the scientific community. In fact, compared to serial kinematics, the possibilities in parallel and hybrid kinematics are lots broader and still unexploited (19).
Dr. Merlet (49) insists on the importance of a methodological creation of kinematics and points out the complexity of this problem. All future characteristics of a machine highly depend on the kinematics, its choice is therefore absolutely vital, and this at the very beginning of the development process.
There exist a lot of methods to create kinematics, the thesis of Dr. Helmer (35) provides a very good overview on this subject. However, these methods are mostly applied to very special conditions like small movement amplitudes, orthogonal arrangement of the axes or the use of only traction-compression linkages (like used in Stewart platforms).

4. STATE OF THE ART: HYBRID KINEMATICS FOR MACHINING TASKS

Additionally, they mostly have only 1 criteria for the generation of kinematics, the amount of degree of freedom.

They are restricted in a mathematical way and often do not take into account concrete design considerations like the arrangement of the linear axes or rotational axes, technological limits of the components, or the simplicity of the mechanism.

It is obvious that the further development of more, general and design-oriented methods will find big success in the scientific community and in the industry.

For now, one of the most successful methods is based on *catalogues of kinematics*. The designer uses known complete kinematics[1], sometimes just parts of them, and by *combination-reorientation-modification* adapts them to their specific problem. As an example, Hale (34) proposes a catalogue of kinematics with orthogonally-distributed struts for all possible combinations of rotations and translations (see figure 4.1). Brogardh (11), in figure 4.2, proposes different kinematic chains which could be used to obtain the correct amount and type of DOF on the robot's output.

Despite ongoing efforts it is nowadays still not possible to create automatically[2] perfectly adapted kinematics, the experience of the engineer is still required[3] for this very determining step. *The synthesis of kinematics is considered as a handcraft sometimes assisted by methods, but generally supported by the creativity and intuition of the engineer.* These features belong to the most important characteristics of an engineer and can be cultivated and stimulated. *The synthesis based on catalogues of existing kinematics creates a stimulation that mostly, and very naturally, leads to inventing new, or adapting existing kinematics in order to suit a given problem.*

A lot of inspiring collections of mechanisms can be found in databases on the web, as for example the webpages of Dr. Merlet (51), Dr. Bonev (9) or the Robotool website (2). There also exist several books compiling, in a very broad way, mechanisms which can be adapted to robotics, for example Sclater (66) or Artobolevski (5).

By definition, a hybrid kinematics must contain at least one parallel part. In the following collection the amount of DOF of the parallel part, called *Degree of Parallelism*, will serve to classify the kinematics. If multiple parallel parts are present in a hybrid kinematics, the highest degree of parallelism will be determining.

---

[1] In general, any kind of mechanism that transforms movement can be used and/or adapted.
[2] Based on general methods or algorithms.
[3] Fortunately.

4.1 Kinematics

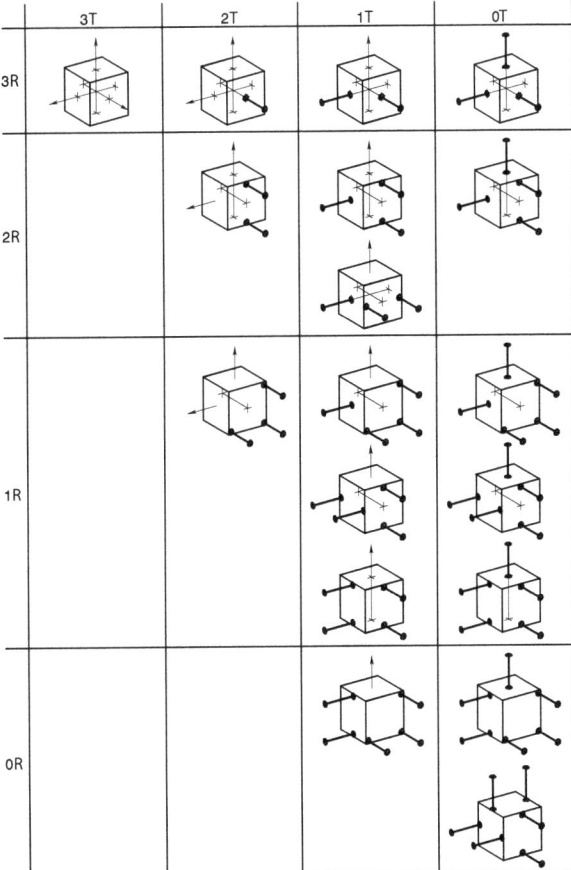

Figure 4.1: Hale (34) proposes a catalogue of kinematics using orthogonally-distributed struts of type SS (balljoint-segment-balljoint) for small-amplitude movements. The actuation is not considered, it can be added in different possible ways (see chapter 6.1). The choice goes from the kinematics with 6 DOF (type Stewart Platform) with 3 translations and 3 rotations (3T3R) to the completely constrained kinematics (0T0R).

4. STATE OF THE ART: HYBRID KINEMATICS FOR MACHINING TASKS

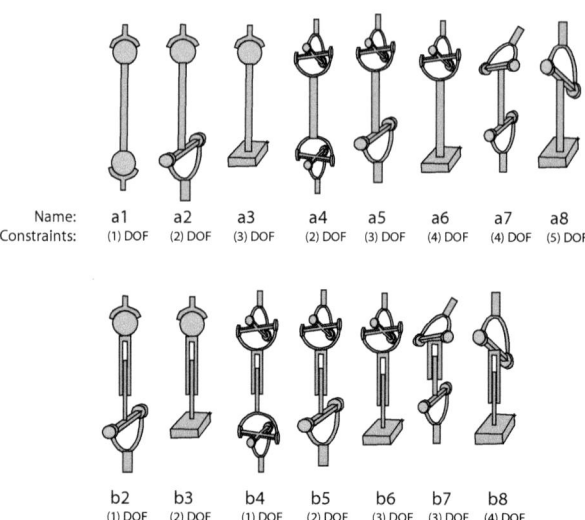

Figure 4.2: Brogardh (11) proposes a catalogue of kinematic chains constraining more or less the robot's output. The amount of constrained DOF (generalized forces GF_i, see equation 1.1) is indicated below the corresponding mechanism.

Note: If the reader wishes to be informed about industrial, *parallel* mechanisms we would like to mention the Robotool website (2) which contains a complete collection of machines.

4.1 Kinematics

4.1.1 Degree of Parallelism = 2

Chiron (www.chiron.de) designed a 5-axes hybrid-kinematic machine tool (62) with a 2 DOF parallel module that generates the translation in the $X - Y$ plane (see figure 4.3). This arrangement of 2 axes is often called *scissor*-kinematics because of its resemblance to the movement of scissors. The actuation of the parallel kinematics is realized by 2 linear motors placed on both sliders of the *scissor*. The third translation Z is added in series. The axes achieve a maximum velocity of $120 \frac{m}{min}$ with an acceleration of $3g$.

The rotations are guaranteed by a serial kinematics and placed on the left hand, a complete 5-side machining[1] is possible.

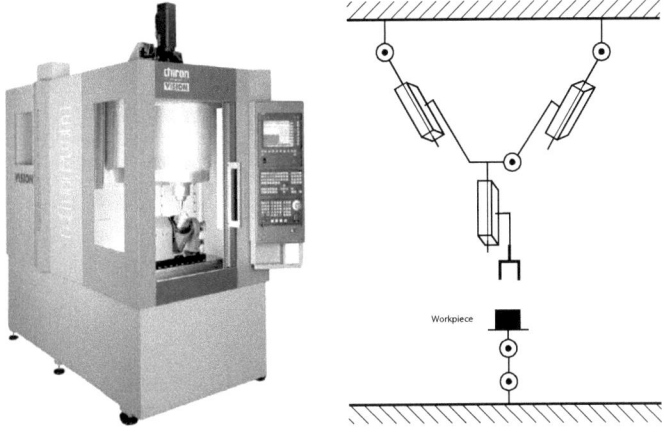

(a) The Vision machining center. (b) Kinematic scheme of the Vision Machine from Chiron.

Figure 4.3: The Vision machine tool by Chiron (www.chiron.de). The kinematics are slightly overconstrained, in order to become isostatic the 2 pivots in the right kinematic chain should be replaced by a spherical joint and a universal joint.

MAG Powertrain (www.mag-powertrain.com) produces a 5-axes machine tool,

---
[1] 5-side machining is often used to declare that the mechanism can reach, and process, 5 sides of a cube while having the tool oriented perpendicularly to each surface. In order to achieve this, the rotation axes θ_x and θ_y need a minimal rotation amplitude of $\pm 90°$.

4. STATE OF THE ART: HYBRID KINEMATICS FOR MACHINING TASKS

the Genius 500, with very similar kinematics. Here again, most of the mechanism is based on serial kinematics except the $X-Y$ translation which is guaranteed by a *scissor* kinematics. The spindle is oriented horizontally and the 3rd translation (in Z) is located on the left hand, as we can see from the figure 4.4. The machines achieves a maximum velocity of $120 \frac{m}{min}$ and an acceleration of $2.4g$.

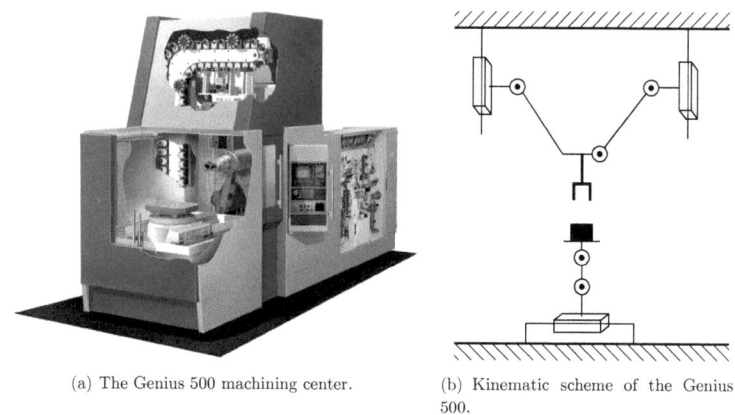

(a) The Genius 500 machining center.

(b) Kinematic scheme of the Genius 500.

Figure 4.4: The Genius 500 machine tool by MAG Powertrain (www.magpowertrain.com). The parallel-kinematic translator is overconstrained.

Kovosvit (www.kovosvit.cz) produces a 5-axes machine tool, the Trijoint 900H, with the same kinematics as the previously shown Genius 500. It is represented in figure 4.5. According to Kovosvit it can exceed $1g$ acceleration and guarantee a stiffness higher than $100 \frac{N}{\mu m}$ in all directions. The first Eigenfrequency is located around $100Hz$.

The **IFW** (Institut für Fertigungstechnik und Werkzeugmaschinen) at Hannover University (www.ifw.uni-hannover.de) developed a 5-axes machining center which differs from the previously shown kinematics. There are still 2, out of the 5 DOF, that are implemented with a parallel kinematics. The 2 parallel axes are moving the serial head on a spherical surface that is defined by the central strut (see kinematic scheme in figure 4.6), thereby generating the $X-Y$ movement.

Mitsubishi Automation (www.mitsubishi-automation.de) also has developed a

4.1 Kinematics

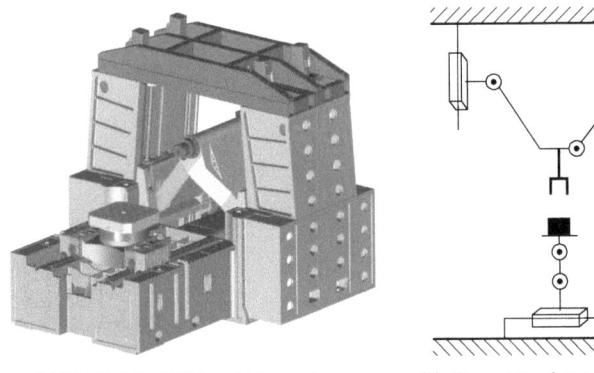

(a) The Trijoint 900H machining center.

(b) Kinematic scheme of the Trijoint 900H.

Figure 4.5: The Trijoint 900H machine tool by Kovosvit (www.kovosvit.cz). Its kinematics are the same as the previously illustrated Genius 500 (see figure 4.4).

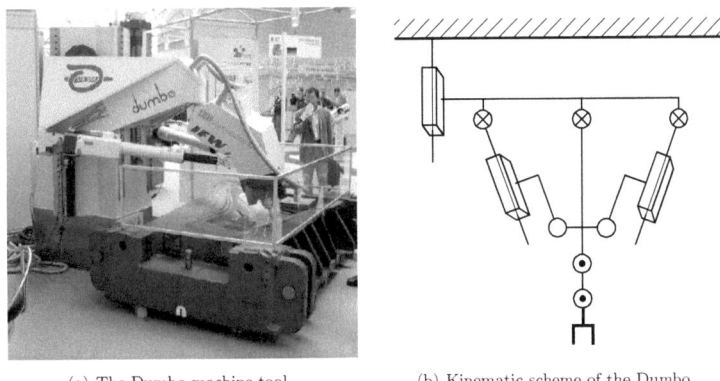

(a) The Dumbo machine tool.

(b) Kinematic scheme of the Dumbo.

Figure 4.6: The Dumbo machine tool by the IFW Hannover (www.ifw.uni-hannover.de). It is designed as a mobile machine tool for the on-site repair of large stamping dies.

hybrid kinematic machine (see figure 4.7). It is often called a *double-SCARA* kinematics because of its obvious familiarity with the well-known SCARA Robot. The robot is used

4. STATE OF THE ART: HYBRID KINEMATICS FOR MACHINING TASKS

for handling tasks like packaging or pick-and-place operations[1]. The robot possesses totally 4 DOF, whereof 2 are implemented as parallel axes[2]. Besides all translations in space, it features the rotation around the vertical axis, the typical movements needed for handling tasks in the horizontal plane. The resulting performances are pick-and-place cycles of $0.5s$ and a repeatability of $\pm 5\mu m$ which is good for handling robots of this size. The robot has a working plane $(X - Y)$, performed by the parallel axes, of $150 \text{x} 105 mm$ for a footprint of $200 \text{x} 160 mm$. The resulting ratio working plane/footprint is very effective.

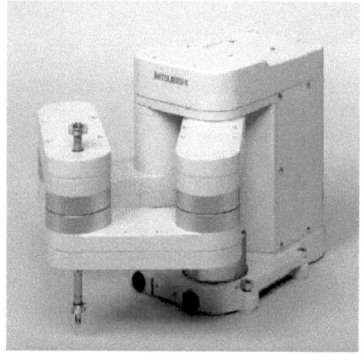 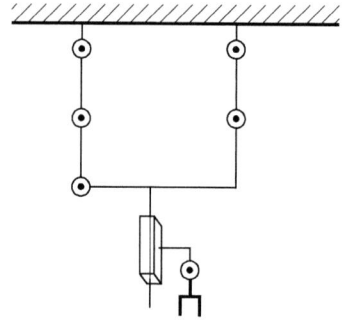

(a) The Mitsubishi RP handling robot.

(b) Kinematic scheme of the Mitsubishi RP. The parallel-kinematic translator is overconstrained. The right arm should be constituted by a balljoint and a universal joint, instead of 2 pivots, in order to become isostatic.

Figure 4.7: The Mitsubishi RP (*"double-SCARA"*) Robot (www.mitsubishi-automation.de).

---

[1] It is therefore not considered as a machine tool. However, its kinematics could be used therefore.
[2] *Scissor* kinematics with a rotative actuation

4.1 Kinematics

4.1.2 Degree of Parallelism = 3

Based on the experience made with the Tricept (figure 1.8) robot, K.-E. Neumann developed a new hybrid kinematics, the Exechon (56). This hybrid 5-axes kinematics is represented in figure 4.8. The modifications and enhancements of this kinematics are based on the Tricept, but the main motivation for this design was the elimination of the central, passive strut, which was exposed to flexion and torsion. The way this strut was exposed to certain loads forced the designers to design it large, and thereby making it heavy and bulky.

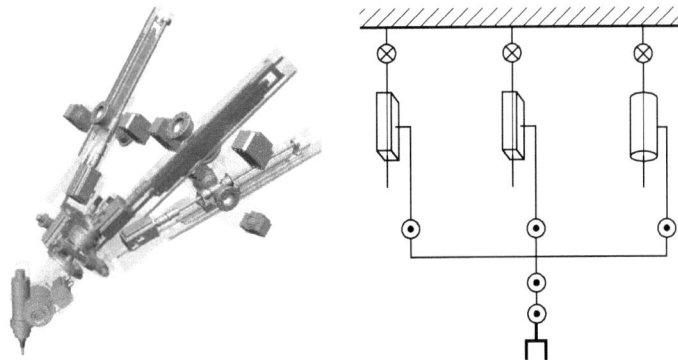

(a) Image of the Exechon machine tool. A 5-axes hybrid kinematics for machining tasks.

(b) Kinematic scheme of the Exechon machine tool. The translator composed of 3 kinematics chains is clearly visible in the upper part of the scheme. The 2 rotations are added in series. The machine, as it is built, is 2-times overconstrained.

Figure 4.8: The Exechon (www.exechon.com).

The Exechon's carrier (or translator) is built up from 3 kinematic chains, whereof 2 are identical. The torsion and flexion constraints – compared to the Tricept – have to be hold by the 3 active struts. The actuation is done on the 3 sliders using ballscrew drives, the position measurement is achieved using linear encoders. The whole drive system is integrated in the struts.

The company **Serramec** (www.serra-aeronautics.com) is selling[1] an Exechon under

---
[1] The business of the company Exechon (www.exechon.com) is built up around the selling of licences

4. STATE OF THE ART: HYBRID KINEMATICS FOR MACHINING TASKS

the name Serramec HK 700. Its characteristics are a stiffness of 100 $\frac{N}{\mu m}$ in Z – along the tripods height – and 50 $\frac{N}{\mu m}$ in X and Y, a repeatability of 10 μm and acceleration of $1g$ in Z and $3g$ in X and Y. The machine is designed for rather big workpieces of 2000x1500x600mm $(X/Y/Z)$ and therefore exhibits a moving mass of 950 kg.

DS Technologie (www.ds-technologie.de) developed a very successful 5-axes hybrid machine tool, the ECOSPEED (37)(38). Its application is the machining of very large, but shallow, aircraft parts (wing and body components). The choice made was to use 2 serial axes (X/Y) to provide large strokes for the displacement along the large workpieces. The 2 rotations, as well as the Z translation, were implemented as a parallel-kinematic module to ensure high dynamics and stiffness. This 3-DOF head is called the Sprint Z3 and is illustrated in figure 4.9. The stiffness of the machine amounts 50 $\frac{N}{\mu m}$ in X/Y and 250 $\frac{N}{\mu m}$ in Z. The machine can achieve a acceleration of $1g$.

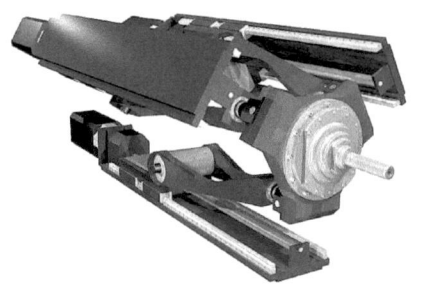 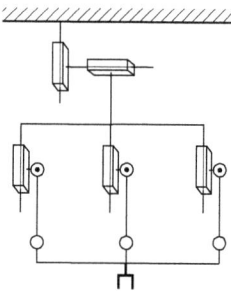

(a) Image of the Sprint Z3 Head. This module has 3 DOF, 1 translation and 2 rotations. It is used in combination with 2 serial axes for translations, by this constituting the ECOSPEED 5-axes machining center.

(b) Kinematic scheme of the whole machine, the ECOSPEED.

Figure 4.9: The Sprint Z3 head from DS Technologie (www.ds-technologie.de). This head is part of a complete 5-axes machining center.

The actuation of the Sprint Z3 head is done on 3 collinear sliders using ballscrew drives, and the position measurement is done with linear encoders. The Z3 head can carry spindles with up to $80kW$ power. Between the market launch in 2000 and the year 2003 DS Technologie sold 18 of these machines (1). Despite its slightly higher

---

to manufacture machine tools with Exechon kinematics

4.1 Kinematics

price than similar machines, it could convince the customers because of its higher productivity and better surface finish[1] of the machined parts.

M Torres (www.mtorres.es) developed in collaboration with **FATRONIK-Tecnalia** (www.fatronik.com) a very similar machine tool concept, the Space 5H (21). The global kinematics are very close to the ECOSPEED, except the parallel-kinematic head which presents some small differences. In fact its DOF are the same, but generated differently by using 6 struts (balljoint-segment-universal joint). 2 struts are mounted pairwise on a same sliding point (see figure 4.10). The difference lies in the way the struts are exposed to loads, as they can only transmit traction-compression loads. This concept has a higher, theoretical stiffness than the concept used for the ECOSPEED Z3 head.

 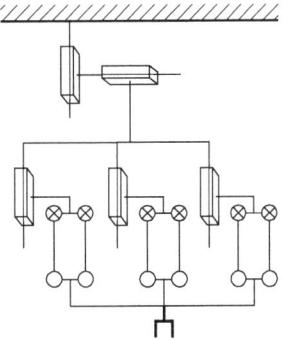

(a) Image of the Hermes head. Like the Sprint Z3 presented above, this module is combined with a serial translator.

(b) The Hermes head is mounted on two serial axes for the translations in a plane.

Figure 4.10: The Hermes parallel-kinematic tilting head from FATRONIK-Tecnalia (www.fatronik.com).

FATRONIK-Tecnalia (www.fatronik.com), a company that is very active in the domain of parallel and hybrid mechanisms, developed another machine called VERNE (47)(41). It is represented in figure 4.11. The right hand forms the parallel-kinematic module – a DELTA kinematics – which accomplishes the translations in space ($X/Y/Z$). The left hand is a serial-kinematic tilting table with 2 rotational DOF.

---

[1] Mostly thanks to a high stiffness and high 1st eigenfrequency.

4. STATE OF THE ART: HYBRID KINEMATICS FOR MACHINING TASKS

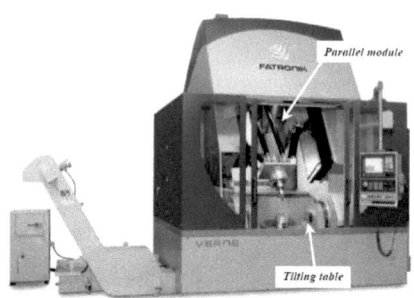
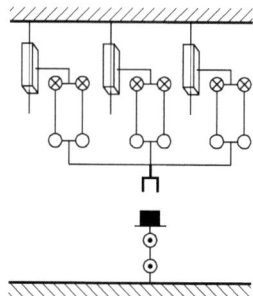

(a) The Verne machine tool.

(b) Kinematics of the Verne machine tool.

Figure 4.11: The Verne 5-axes machine tool by FATRONIK-Tecnalia (www.fatronik.com).

Reichenbacher GmbH (www.reichenbacher.com), a company specialized in the machining of wooden parts, also developed a machine concept based on a DELTA kinematics, the PEGASUS (23). The 5-axes version illustrated in the figure 4.12 constitutes a hybrid kinematics. Unlike the VERNE, the PEGASUS accumulates all DOF on the right hand, the main reason being the difficulty to move the large wooden workpieces. The machine reaches advance-speeds of 120 $\frac{m}{min}$ with accelerations up to $1g$.

4.1 Kinematics

 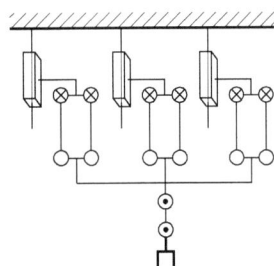

(a) The Pegasus 5-axes version with double-tilting head.

(b) Kinematic scheme of the 5-axes hybrid kinematics.

Figure 4.12: The Pegasus machine tool by Reichenbacher GmbH (www.reichenbacher.com). This machines presents a very similar kinematics to the Verne machine illustrated in figure 4.11.

Figure 4.13: The Pegasus basic concept with 3 DOF in translation. This version is a purely parallel kinematics.

As we have seen with the last 2 examples, the Delta kinematics (14)(15) inspired a lot of designers and lead to the development of an impressive amount of machines. For this reason 2 additional mechanisms, both nor hybrid kinematics nor machine tools, will be cited here:

ABB Robotics (www.abb.ch) developed a 4-axes pick-and-place parallel robot based on the Delta kinematics, the IRB 360 Flexpicker (see figure 4.14). A forth DOF is added to the 3 translations in space (a variant which is included in the Delta patent

4. STATE OF THE ART: HYBRID KINEMATICS FOR MACHINING TASKS

(14)), allowing the robot an infinite rotation around the vertical axis. The presented machine achieves velocities of 600 $\frac{m}{min}$ with accelerations up to $15g$.

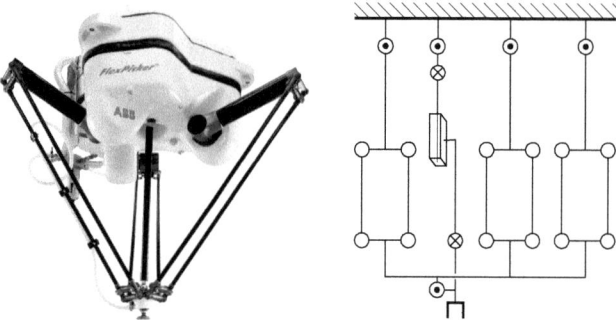

(a) The ABB 360 Flexpicker, a 4-axes, hybrid kinematics for high-speed pick-and-place applications.

(b) Kinematic scheme of the Flexpicker.

Figure 4.14: The ABB IRB 360 Flexpicker, based on a Delta4 kinematics.

Adept (www.adept.com) commercializes a Par4 mechanism (52) under the name Quattro (figure 4.15). The mechanism is intended for manipulation tasks like packaging. It performs movements in x, y, z, θ_z with high rotation amplitude on the rotative axis θ_z and is considered, by Adept, as the world's fastest industrial packaging robot, allowing movements speeds up to 600$\frac{m}{min}$ with accelerations up to $15g^1$. The rotative movement is achieved by the relative displacement of the 2 end-effector parts and amplified by a means of a transmission.

---

[1]The datasheets of the IRB 360 Flexpicker (source www.abb.com) and Quattro (source www.adept.com) indicate the same velocities and accelerations.

44

4.1 Kinematics

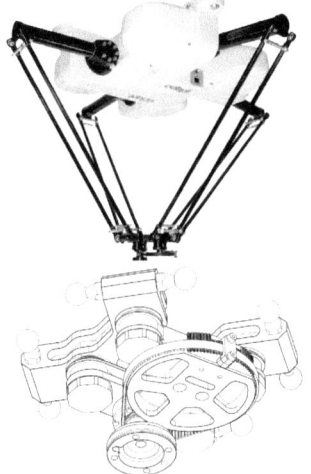
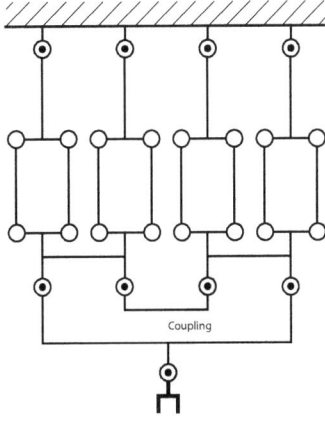

(a) The Par4 mechanism, commercialized under the name Quattro by Adept (www.adept.com), and a detailed illustration of the end-effector.

(b) Kinematic scheme of the Par4 mechanism. The amplification system is outlined but not truthfully represented. The 4 pivot joints at on the end-effector slightly overconstrain the mechanism.

Figure 4.15: The 4-axes Par4 mechanism (52) $(x, y, z, \theta_z = \pm 180°)$ using an amplification system to increase the rotation of the end-effector.

4. STATE OF THE ART: HYBRID KINEMATICS FOR MACHINING TASKS

4.1.3 Degree of Parallelism = 4

Hybrid kinematics which possess a parallel module with 4 DOF are rare. One example is the HITA STT (20) developed at the **LSRO (lsro.epfl.ch)** in collaboration with **Willemin Macodel** (www.willemin-macodel.ch). It is represented in figure 4.16. Additional to all three translations, there is also 1 rotation that is achieved by the parallel kinematic module. As already mentioned in section 3.2, this machine is one of the only – parallel kinematics[1] – that can perform angles higher than 90° without affecting the global stiffness of the machine. The HITA can perform 5-axes machining, achieves velocities of 120 $\frac{m}{min}$ with accelerations of $5g$. Its stiffness amounts 12 $\frac{N}{\mu m}$ and its 1st Eigenfrequency lies at $120Hz$ (71)(70).

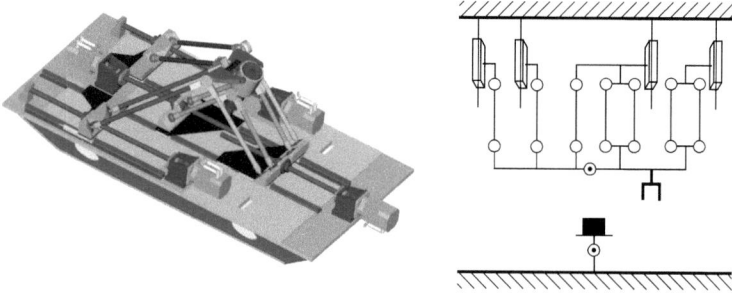

(a) Illustration of the Hita STT

(b) Kinematic scheme of the Hita STT. The parallel-kinematic module on the right hand possesses 4 DOF whereas the left hand is constituted by a single rotational axis.

Figure 4.16: The Hita STT hybrid machine tool developed at the LSRO (lsro.epfl.ch).

---
[1]Considering only the 4-DOF parallel mechanism.

4.2 Elements

A machine is a compound of different elements all having different functionalities and influences on the performance of the machine. As these elements greatly define the final characteristics of the machine, they should therefore be effective and adapted to their task. This chapter will investigate the major elements of a machine: *the joints and the drive systems*.

The joints greatly influence the characteristics of the whole machine as they enable its mobility, define its stiffness[1], highly influence its precision/repeatability and its reliability. The first part of this section will analyze the state of the art in joint development.

The drive systems allow and define the machine's dynamics (velocities and accelerations) and ability to exert forces at the tooltip. The second part of this chapter will investigate the mostly used drive systems for parallel/hybrid kinematics in the domain of machine tools.

There exist different technologies to achieve movement for a joint. The most common technology are interface-based joints, where the movement is achieved with multiple elements interacting either by gliding or by rolling on each other. More recent technologies applied to robotics, like flexure-based joints, have appeared in the last years. The following table enumerates the differences:

Interface-based joints are the classical joints composed of multiple parts interacting with each other via a mechanical interface. The interaction can take place by rolling (e.g. ball/roller bearings) or gliding/sliding (plain bearings). The difficulty of these joints rests in handling the interface (its general shape, surface finish, manufacturing, dimensioning and modeling).

Flexure based joints perform movement by the localized and elastic (sometimes superelastic) deformation of a part. A joint is therefore constituted of one single part. In certain cases multiple joints and segments are regrouped on 1 part which is then called a *monolithic* structure.

Mixed technology joints are a compound of the two above-mentioned technologies.

---

[1] As they normally are the weakest elements in the kinematic chains.

4. STATE OF THE ART: HYBRID KINEMATICS FOR MACHINING TASKS

4.2.1 Interface-based Joints

Simple joints that execute only 1 DOF – like pivots or sliders – have seldom been subject of profound research in parallel/hybrid robotics. In reality, the performance of these simple joints is already very well handled. As they are used in both domains, parallel kinematics and serial kinematics, as well as in a lot of other domains that are not directly linked to robotics, there has been a lot of development done so far.
Joints with 2 or higher mobilities, like spherical joints or universal joints, are typical elements of parallel and hybrid mechanisms. They are used in great quantities, resulting from the natural, large sum of mobilities of parallel/hybrid kinematics.

INA (www.ina.com) proposes a solution for both spherical- and universal-joint, as illustrated in figure 4.17. These components are used in a lot of industrialized hybrid machines like the Verne (figure 4.11), the Hermes (figure 4.10), the Z3 (figure 4.9) and other fully-parallel kinematic machines like the Starrag Heckert SKM (65) (see figure 4.18). These joints represent the state of the art of commercially available joints, specially designed for parallel/hybrid kinematics.

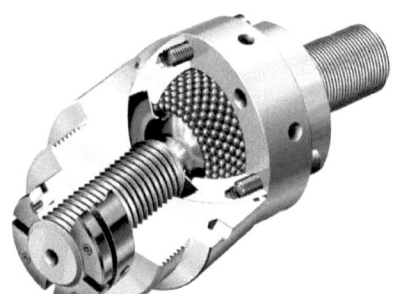

(a) The universal joint from INA (type GLK 2 or GLK 3), based on a serial and concentric arrangement of ball bearing axes.

(b) The balljoint from INA (type GLK).

Figure 4.17: The 2 types of joints INA (www.ina.com), proposed for parallel-kinematic machines.

The 2 tables 4.1 and 4.2 summarize the main characteristics of these joints, depend-

4.2 Elements

ing on their size (small or big version).

Table 4.1: Characteristics of the INA joints, **small versions**

Characteristic:	GLK Balljoint	GLK 2 Universal joint	GLK 3 Balljoint[a]
Mass [kg]	2.3	0.95	1
Swivel angles [°]	±20° ±20° ±360°	±45° ±90°	±45° ±90° ±360°
Stiffness [$\frac{N}{\mu m}$]	280	50	50
Required space [mm]	⌀70x70	⌀82x108	⌀82x108
Price [CHF] approx.	1400.-	NA	1500.-

[a]The GLK 3 Balljoint is a combination of 3 concentric pivots where all axes intersect in 1 point

Table 4.2: Characteristics of the INA joints, **big versions**

Characteristic:	GLK Balljoint	GLK 2 Universal joint	GLK 3 Balljoint
Mass [kg]	4.5	7	7
Swivel angles [°]	±20° ±20° ±360°	±45° ±90°	±45° ±90° ±360°
Stiffness [$\frac{N}{\mu m}$]	350	450	450
Required space [mm]	⌀90x94	85x135x128	85x135x128

As shortly indicated in the tables 4.1 and 4.2, the GLK 3 joint achieves its movement by combining 3 pivots in series. The advantage of this combination are the large swiveling angles, compared to a balljoint composed of a ball and a socket. However, there are two major drawbacks, first, the serial arrangement of joints which complicates the mechanics and strongly influences the global stiffness (serial arrangement of compliances), and second, the possible[2] singularity when 2 out of the 3 axes become collinear. The orientation of the joint and its swiveling range become crucial and have to be analyzed systematically.

Although there are only few commercially available joints that are suited to parallel/hybrid mechanisms, there has been a lot of development in this domain. Schnyder (64) presents the development of single and double spherical joints for the HITA STT prototype. The single balljoint is represented in figure 4.19. The joint presents two

[2]The INA balljoint does not present this problem, but in general the arrangement of 3 concentric axes can lead to it.

4. STATE OF THE ART: HYBRID KINEMATICS FOR MACHINING TASKS

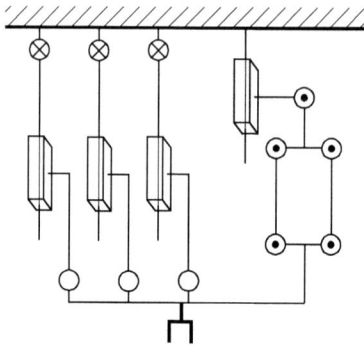

(a) The Starrag Heckert SKM machine tool. The 3 spherical joints that are close to the spindle are from INA, as well as the 3 active telescopic struts.

(b) The kinematic scheme of the Starrag Heckert SKM. The parallelogram formed by the 4 pivots is highly overconstrained. This can be prevented by allowing some precisely defined and localized parts to bend, or, by substituting 2 pivots by a universal joint and a spherical joint.

Figure 4.18: The Starrag Heckert SKM (www.starragheckert.com). An example of a fully-parallel 3-axes machine tool using INA standard components.

distinct spherical contact regions, one with a much smaller radius on which the preload (2000 N) is applied. This considerably reduces the friction moment. The spherical surface with the bigger radius serves as the geometrical and mechanical reference of the joint, it defines the stiffness of the joint. This joint has a stiffness of 500 $\frac{N}{\mu m}$ and can achieve angles up to $\pm\,15°$.

4.2.2 Flexure-based Joints

Flexure-based joints, or commonly called **flexures**, are based on the elastic deformation of precisely defined areas of a part. This deformation is always either bending or torsion, and therefore a single joint will always only produce rotative movements. More complex movements can be achieved by combining several flexures. The use of this type of joints leads to a multitude of advantages, as presented in table 4.3.

Henein (36) wrote a very complete thesis about the use and the dimensioning of

4.2 Elements

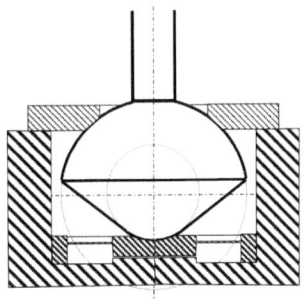

(a) Image of the single joint. (b) Section through the joint.

Figure 4.19: The spherical joint used in the Hita STT machine tool.

Table 4.3: Advantages and disadvantages of flexures

Advantages	Disadvantages
• No blacklash • High stiffness • No friction • No wear • High precision movement because of the above advantages • Less sensitive to polluting elements/substances	• Limited range of movement due to the elastic limits of the material • Restoring force (elasticity of part) • Material fatigue • Geometric modeling: The instantaneous center of rotation is depending on the performed angle. Therefore, modeling is difficult.

flexures. He determines the movement amplitudes as a function of the flexure's dimensions, the corresponding stiffness in all directions, and the maximum internal constraints which shouldn't be transgressed in order to guarantee a certain lifetime of the joint. Pernette (57), Helmer (35) and Bacher (7) made use of flexures to create complete mechanisms for ultra-high-precision applications. One of these mechanisms, the Trib-

4. STATE OF THE ART: HYBRID KINEMATICS FOR MACHINING TASKS

ias, is presented in figure 4.20.

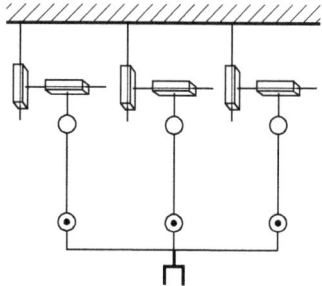

(a) The Tribias robot (57). The pivot joints close to the robot output are realized by the use of special high-stroke flexures.

(b) The simplified kinematic scheme of the Tribias. Because of the use of rotative flexure joints, and the need for high rotation amplitudes, the real kinematic scheme is lots more complicated. This kinematic scheme represents a simplified version and uses sliders.

Figure 4.20: The Tribias, a hybrid 6 DOF robot for high-precision assembly applications.

Despite their biggest drawback, the limited range of movement, their implementation is highly recommended for applications in the domain of machine tools. This, because of their mechanical performance (in terms of precision and stiffness) and their reliability in polluted environments. Their maximum strokes can be determined by using the formulary given by Henein (36). Their stiffness decline with augmenting strokes, and this results in a realistic application interval from $0°$ to $\pm 10°$.

4.2 Elements

4.2.3 Mixed Technology Joints

Mixed technology joints are quite recent and only few developments have been subject to scientific literature. This type of joint combines the advantages of flexure-based joints (no friction, high precision) and interface-based rolling joints (high load, high stroke). Figure 4.21 shows a 1-DOF joint (pivot) developed at the EPFL (4). The two rolling surfaces provide stiffness in traction/compression and bending (around θ_y). The crossed, flexible strips transmit the preload and prevent the joint from twisting (around θ_z). Their arrangement compensate the typical restoring force (due to elasticity) and dictate a perfect rolling movement between the 2 parts of the joint, this way, no slipping is possible. Their mode of operation can be closely compared to a human knee.

Figure 4.21: A mixed technology joint from the EPFL, developed by Fracheboud and Allemand (4).

Cannon (12) developed several types of joints. The basic principle, the CORE joint, is shown in figure 4.22. The joint is built up by multiple, identic monolithic parts arranged on different layers.

4. STATE OF THE ART: HYBRID KINEMATICS FOR MACHINING TASKS

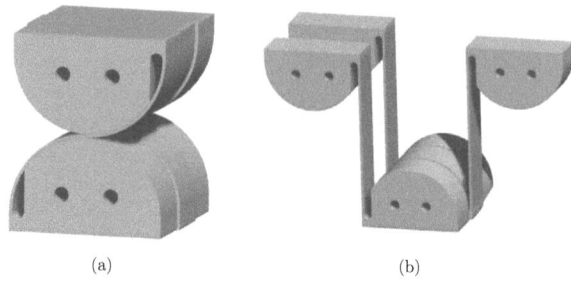

Figure 4.22: The Core (compliant rolling-contact element) joint (12).

4.2.4 Drive systems

This chapter will focus on the different methods of *actuating the parallel-kinematic module* of a hybrid machine tool. As there are multiple joints in a parallel kinematic chain, the designer has to choose which DOF he will actuate. For the serial kinematic parts however, this question does not occur, the joint that has to be actuated is clearly defined.

The parallel-kinematic part has 2 common ways to be actuated. Its kinematic chains are either actuated by **moving the base-point** or by **varying a strut-length**. Both of these methods are illustrated in figures 4.23 and 4.24. Beyond the machine tools there exist other methods for actuation (e.g. base-point moving on a rotative lever, typically used in pick and place applications) but they are not adapted to the stiffness requirements that are common practice in the machine tool industry.

Actuating a DOF in a parallel-kinematic chain is mostly made on a joint that provides translation (e.g. slider). The reason is that this drive system can be picked up from traditional serial machines tools.

Moving base-point: This method consists in translating the base-point along a mostly straight trajectory (slider). The following structure segment can therefore be fixed in length. The actuation is done using any type of linear actuator: Ballscrew drive, direct-drive linear motor, pneumatic piston etc.

Varying strut-length: This method consists in actuating a slider in-between two

4.2 Elements

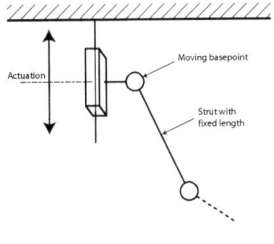

(a) Moving base-point principle.

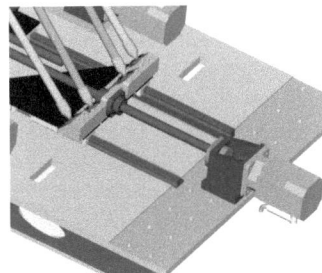

(b) Moving base-point system of the Hita STT. A ballscrew system actuates the slider which supports the base-point (in this case some balljoints).

Figure 4.23: The moving base-point drive system.

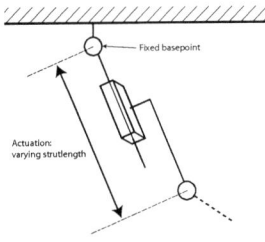

(a) Variable strut-length principle.

(b) Variable strut-length drive system on a flight simulator from Sikorsky (www.sikorsky.com) at the FlightSafety (www.flightsafety.com) facilities. The 6 identic struts synthesize a parallel kinematics called **Stewart platform**. One of these legs is represented in the kinematic scheme on the left.

Figure 4.24: The variable strut-length drive system.

rotative joints. The resulting system is often called a telescopic actuator. The base-point is fixed.

4. STATE OF THE ART: HYBRID KINEMATICS FOR MACHINING TASKS

Nearly all motor/driving technologies have been investigated and tested on parallel/hybrid mechanism (electric servo-drives in all variants, hydraulic struts, piezo-actuators, pneumatic struts, pneumatic muscles etc.). Today's most represented setups consist of industrial, rotative servo-drives combined with ballscrews and linear bearings. Already very well known from serial machine tools, this technology can be rapidly adapted to parallel/hybrid machines. A clear tendency today is the use of direct-drive linear motors. Their advantages are extremely interesting, as they possess less inertia for a same power, a higher stiffness and less wear because of the absence of multiple mechanical parts (ballscrew, nut, eventual gearbox). The downside of this technology is the complexity of an effective protection system, as in a linear motor the rotor and stator are unwinded and can attract ferromagnetic particles.

4.3 Summary and Conclusion of the Chapter

This chapter provides an overview of hybrid mechanisms, their different elements, and the different technologies they are made of. All mechanisms were presented with their kinematics scheme. Besides completing the catalogue of kinematics, this state of the art will be valuable for the following chapters, as there will be several references on these mechanisms. An analysis of the mentioned hybrid kinematics, and of some representative parallel kinematics, will be made in chapter 5.

Commercially available high-mobility joints are rare to date, compared to the widespread sliders of serial kinematics. For a lot of machines proprietary joint designs are developed. This introduces additional risks when developing new machines and requires a rigorous testing. Furthermore, the companies which emerge in the field of parallel/hybrid kinematics will not posses the necessary knowhow and will first need to acquire it. Chapter 6 will therefore introduce a novel and well-performing joint design including its modeling.

5

Hybrid Kinematics: A Trade-Off between different Characteristics

We already mentioned the advantages that parallel kinematics have over serial kinematics (chapter 3), but despite this, the few disadvantages remain discouraging for the machine industry. The smaller linear and angular working volume (for a given foot print) seems most important in the decision about switching to parallel kinematic technology. The industries slowly realize and approve the advantages of parallel kinematics, but they perceive them as too radically focused on certain characteristics. Additionally, there are often intimidated by their complexity which leads to doubts about their reliability.

In the machine-tool sector the typical axes requirements are 5-axes $(x, y, z$ and $\theta_x, \theta_y)$ and 5-side machining, which implies angles up to $\pm 90°$ on the rotative axes. Whereas the translation strokes could be increased by scaling the whole mechanism, it is much more difficult to increase the angular strokes. Only few parallel mechanisms can bypass this disadvantage by using ingenious kinematics, gear transmissions or redundancies (Hita STT (71), Alpha5 (59), Par4 (53), Orthoglide 5-axes (13)(44)(see figure D.10 in appendix), but only with a significant increase of the mechanisms' complexity.

In order to illustrate this increase in complexity, we will introduce a concept:

The **Mobility Inefficiency (MI)** ε will be defined as the ratio between the total amount of DOF of a kinematics[1] and the effective DOF[2] of the output:

---

[1] The sum of mobilities of all joints
[2] Redundant mobilities are not taken into account.

5. HYBRID KINEMATICS: A TRADE-OFF BETWEEN DIFFERENT CHARACTERISTICS

$$\varepsilon = \frac{\sum_i n_i}{n_{\text{output}}} \geq 1 \tag{5.1}$$

where n_i is the mobility of joint i and n_{output} is the effective mobility of the output. The following categorization results from this definition:

- **Serial kinematics** $\varepsilon = 1$. Every mobility of a joint directly adds a mobility to the output and is therefore necessarily actuated.

- **Parallel and hybrid kinematics** $\varepsilon > 1$. The sum of all mobilities is higher than the mobility of the output, resulting in a certain amount of unactuated (passive) joints.

Direct consequences of a *high mobility inefficiency* are:

- Mechanical: Increasing amount of elements, interfaces.

- Modeling/Control: Increasing amount of geometric parameters, singularities, possible self-collisions, increase of the complexity of the mathematical/physical models.

- Reliability/Performance: Increased possibilities of failures, increased risk of poor performance.

Parallel kinematics do not have the same level of technical maturity as serial kinematics yet. Therefore, developing architectures with high mobility inefficiency might be risky, all the more if the used joint designs are completely new.

Hybrid kinematics can, as we will see, present a compromise. They can increase the workspace, especially the rotative strokes, though theoretically loosing stiffness and dynamics, and additionally, they can decrease the mobility inefficiency. This can lead to simpler designs, which are less radical and superior to fully-parallel mechanisms.

Side note: The fully-parallel Delta robot for packaging tasks, in its 3 DOF version, overcomes this difficulty by using an original system for its spherical joints (see kinematic scheme in section 1.2.3, and the illustration in figure 5.1). Two spherical joints share a common preloading spring. The spring is fixed on the each tube, and this in such a way that internal DOF is partially constrained. The sockets are pressed on the spherical extension, guaranteeing a backlash-free and preloaded joint. The system is easily mountable and also serves as mechanical fuse in case of a collision, and at last, it performs 6 mobilities totally.

The following sections will illustrate possible effects of hybrid kinematics.

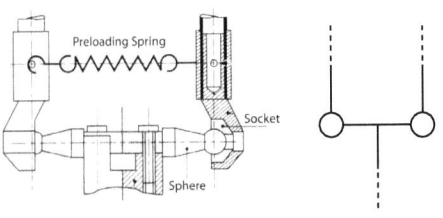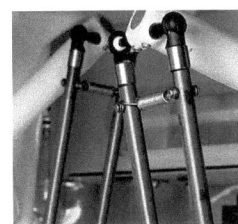

(a) This very simple system accomplishes the 6 mobilities of the two spherical joints.

(b) Equivalent kinematic scheme.

(c) Picture of the Delta robot for packaging tasks, sold over 4000 times.

Figure 5.1: The double spherical joint of the Delta robot. This simple system, developed for the Delta robot, is now commonly used in parallel robotics.

5.1 Reducing the Complexity of the Kinematics

This section compares the complexity of different kinematics, all designed (or intended to be designed) for machining tasks. Every cited example has 5 DOF, the most common axes arrangement for this type of tasks. The complexity will be compared by means of the *Mobility Inefficiency*. Serial, parallel and hybrid kinematics are represented and classified depending on their *Degree of Parallelism* (fully-serial kinematics = 1, fully-parallel = 5, hybrid = 2...4)

Figure 5.2 illustrates the *Mobility Inefficiency* with respect to their *Degree of Parallelism*.

Comments: Two redundant axes were considered as one. The Stewart Platform was added because it is a well represented parallel-kinematic machine for machining, although it is a 6-axes kinematics. The mobility inefficiency of the Stewart platform was computed for 5 DOF of the output, since the 6th axis is not relevant for machining tasks. In order to compare the machines on an equal base, some kinematics were adapted and made isostatic by the author's discretion.

The mechanisms proposed by Feng Gao (27), the Orthoglide 5-axes (13), the Metrom machine (www.metrom.com) and the Omega kinematics developed at EPFL (18) are illustrated in the appendix D.2, as they were not presented in this work yet.

5. HYBRID KINEMATICS: A TRADE-OFF BETWEEN DIFFERENT CHARACTERISTICS

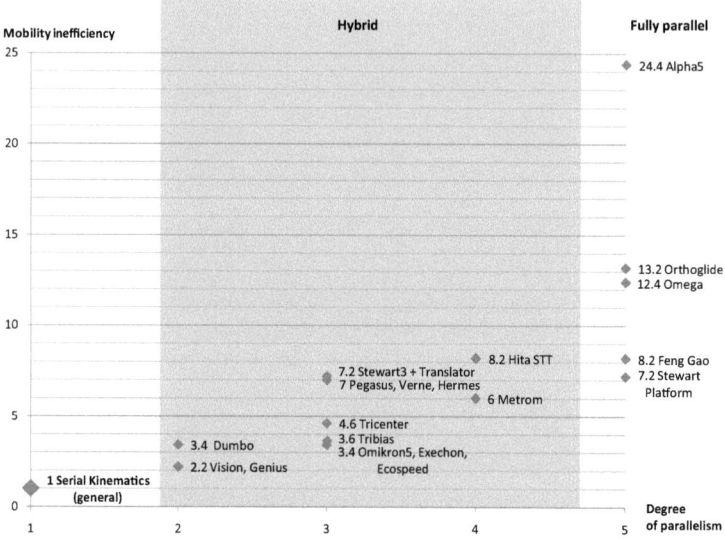

Figure 5.2: Mobility inefficiency of different 5-axes machine tools. The kinematics which were not illustrated yet can be found at following places: *Omikron5*, section 9.3 figure 9.24/ *Stewart3+translator*, section 9.1 figure 9.3/ *Metrom*, appendix D.2 figure D.11/ *Orthoglide (5-axes version)*, appendix D.2 figure D.10/ *Omega*, appendix D.2 figure D.9/ *Feng Gao*, appendix D.2 figure D.12/ *Stewart*, section 4.2.4 figure 4.24.

Several tendencies can be pointed out:

- The higher the degree of parallelism, the more complicated the mechanism becomes: The extreme values are: **Mobility inefficiency=1** for serial kinematics, and, **Mobility inefficiency=24.4** for the fully-parallel Alpha5[1]. In order to generate 1 mobility of the output the Alpha5 requires 24.4 joint mobilities! The complexity of the mechanisms increases nearly linearly with the degree of parallelism. Hybrid kinematics reduce the mobility inefficiency.

- Stewart platforms or variants of 6-strut kinematics (with kinematic chains of

[1] Of course, a more complicated kinematics could still be invented.

type: UPS) are located at a mobility inefficiency of 6-8, even for a high degree of parallelism. Their struts only transmit traction/compression efforts.

- 3-strut kinematics (with kinematic chains of type: RPS) like the Omikron5, Exechon, Ecospeed perform well at a mobility inefficiency of 3.4. Their struts however are loaded by bending moments, therefore, their mechanical design is more critical.

Overconstraining the kinematics can slightly reduce the mobility inefficiency and simplify the design. The Exechon makes use of this concept (56). However, overconstraining kinematics for precision applications (machine tool) is not advised (68)(34) as it can create varying internal constraints and distort precise geometries. Furthermore, no closed-form mathematic modeling is achievable.

In general, a reduction of the mobility inefficiency can be achieved by increasing the motion constraints of each kinematic chain and by reducing the amount of kinematic chains, thus, tending toward serial kinematics. For example, in a Delta robot, switching from spatial double-bars to a single-bar (with 2 universal joints) reduces the inefficiency but adds a torsion constraint to the single bar.

On a kinematic and structural level, a trade-off is made between potential stiffness and complexity. On the level of machine elements however, a reduction of joints and their compliances is achieved.

5.2 Enhancing Movement Amplitudes

If high rotative strokes are needed, typically $\theta_x = \pm 90°, \theta_y = \pm 90°$ for machining tasks, the situation changes as several kinematics are eliminated. Figure 5.3 was modified to contain only kinematics with high rotative strokes.

Again, several tendencies can be pointed out:

- The fully-parallel Stewart Platform and its variants are excluded because they cannot reach the required angles. The lowest mobility inefficiency for fully-parallel kinematics increased to a value around 13 (Orthoglide and Omega).

- There do not exist a lot of fully-parallel mechanisms which can guarantee 5-side machining. The existing few have a quite impressive and sophisticated kinematics.

- A very high increase of the mobility inefficiency can be noticed when switching the degree of parallelism from 4 to 5. Fully-parallel mechanisms are extremely complex.

5. HYBRID KINEMATICS: A TRADE-OFF BETWEEN DIFFERENT CHARACTERISTICS

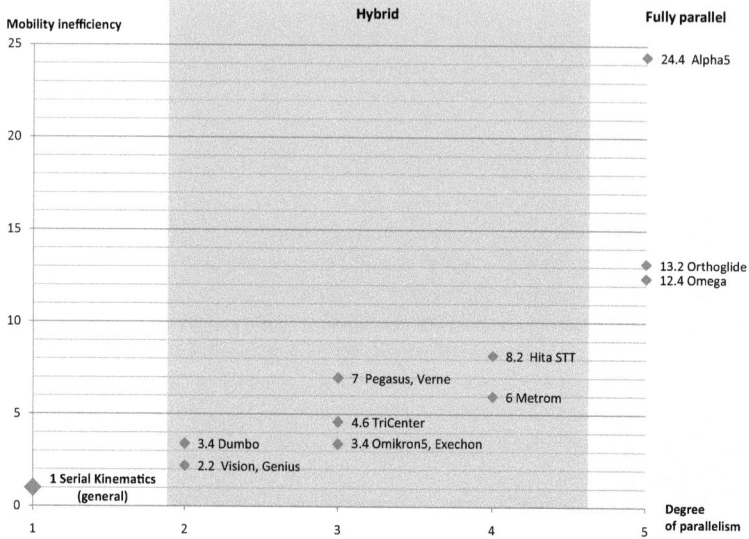

Figure 5.3: Mobility inefficiency of different 5-axes machine tools which achieve high rotation amplitudes.

The existing fully-parallel 5-axes mechanisms are highly complicated and seem not adapted to machining tasks. In fact, their great amount of joints might either compromise the stiffness, or, if the joints have a robust and stiff design, they may drastically decrease the dynamics and eigenfrequency of the mechanism. The higher structural stiffness of parallel machines, an often cited advantage, could be lost because of the high amount of joint compliances. Therefore, insisting on the use of parallel kinematics, when their architecture is complex, can be a wrong choice.

5.3 Conclusion of the Chapter

The mechanisms with a degree of parallelism of 3 are the most popular. In fact, the distribution of 3 parallel et 2 serial axes, with the possibility to increase the stroke of

5.3 Conclusion of the Chapter

specific axes by using serial axes, is the most popular. The Hermes (21) and Ecospeed (37) machine concepts use 2 serial translation axes to cover the very wide workspace of aerospace parts. The Verne (47), Pegasus (23), Exechon (56), TriCenter (32) and Omikron5 (see section 9.3) require high rotative strokes and therefore make selective use of 2-axes serial-kinematic wrists.

In general, the flexibility of axis arrangement, the moderate complexity and the resulting reasonable technological risks are making hybrid kinematics a good trade-off for machining tasks. For 5-side machining however, and from a mechanical and economical point of view, it is not yet realistic to use fully-parallel mechanism. Hybrid kinematics (or of course serial kinematics) are, in this special case and to date, the only viable concept.

5. HYBRID KINEMATICS: A TRADE-OFF BETWEEN DIFFERENT CHARACTERISTICS

6

Machine Elements

6.1 Kinematic Chains

Kinematic chains are assemblies of joints and segments which can carry out two functions: Guiding and actuation. This chapter is concerned with these higher-level elements of a mechanism.

The **guiding function** is about constraining the unwanted mobilities in order to avoid free movement. Every mechanism having less than 6 mobilities needs its kinematic chains to apply movement constraints (or generalized forces GF_i, see section 1.2.1) on the output in order to inhibit the unwanted mobilities.

According to Gruebler, and applied to kinematic chains, we can state the following equation:

$$Mo_{output} = 6 - \sum_i GF_i = 6 - \sum_i (6 - Mo_i - 6l_i) \qquad (6.1)$$

Mo_{output} are the mobilities of the mechanism, GF_i the amount of movement constraints applied by the i-th kinematic chain, Mo_i the mobilities of the i-th kinematic chain and l_i the amount of mechanical loops in the chain. The necessary constraints can be divided equally on all chains, but not necessarily, any distribution can be chosen. A kinematic chain which does not constrain the output ($GF_i = 0$) is called a *non-constraining kinematic chain*[1].

---

[1] Figure 6.1, right side, illustrates 1 possible variant.

6. MACHINE ELEMENTS

Equation 6.1 can provide a simple synthesis[1] method for kinematics. Knowing the mobilities that the mechanism's end-effector should have, we can compute the amount of needed generalized forces $\sum GF_i$ and distribute them on several kinematic chains. The kinematic chains can then be composed in different variants with different complexity, according to the needed amount of generalized forces. A catalogue of the most simple, but often used, kinematic chain with their corresponding amount of constraints is provided by Brogardh (11) in figure 4.2. An impressive collection can be found in the work of Helmer (35).

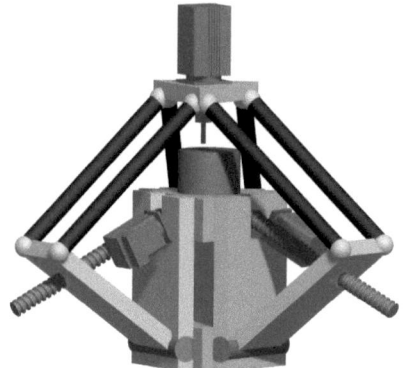 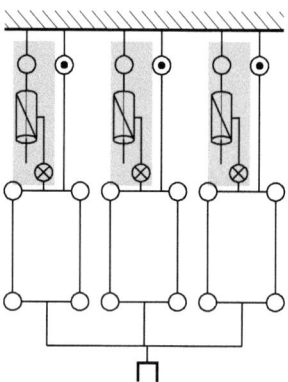

(a) An inverted 3-DOF Delta with augmented kinematics. This concept avoids using linear sliders (which are expensive). The force-loops are kept as short as possible in order to guarantee highest stiffness and precision.

(b) Kinematic scheme. A non-constraining kinematic chain is highlighted. The actuation is carried out on the ballscrew (symbolized by the helical joint) according to the concept of a varying strut-length (see section 4.2.4).

Figure 6.1: An inverted Delta and its augmented kinematics. In order to achieve a complete decoupling of actuation and guiding, the non-constraining chains can be attached directly to the end-effector.

The **actuation function** is imposing movement to a non-constrained DOF of the output. Actuation can happen on any DOF of the mechanisms, but is mostly done on

---
[1] On the other hand, it can represent an analysis method. Problems of overconstraining can be directly credited to chains without computing the mobility of the whole mechanism. A typical example is the parallelogram-shaped four-bar linkage using 4 pivots (figure 4.18).

joints providing linear movement (sliders, helical joints etc.) because of the easiness of implementation (see section 4.2.4).

If there is no suitable joint in a mechanism, the kinematics can be augmented by adding non-constraining[1] kinematic chains, as illustrated in figure 6.1. Another example is the Starrag Heckert SKM in figure 4.18, mentioned in the state of the art, or the Stewart platform which has 6 mobilities and only consists of non-constraining chains. The additional, non-constraining kinematic chains can be composed of joints which allow a simple implementation of actuators.

Augmenting kinematics is an effective way of "parallelizing" kinematics. An initially serial kinematics providing n mobilities can be augmented with n non-constraining kinematic chains (like the Starrag Heckert or the kinematics in the appendix, illustrated in figures D.1 and D.2) and result in a parallel mechanism.

Furthermore, when using such chains, a *decoupling of guiding and actuation* in separate parts of the machines can be achieved. From a mechanical point of view this is interesting, since the mechanical parts of the single chains will bear less types of efforts[2].

The 2 mechanical elements that constitute a kinematic chain, the joints and the segments (rigid bodies), will be treated in the following sections. This section defined the tasks of kinematic chains and proposed ways to adapt and enhance mechanisms by augmenting the kinematics or by decoupling actuation and guiding.

6.2 Joints

As already mentioned in section 4.2.1, the main focus in joint development for hybrid mechanisms lies on high-mobility joints (spherical- and universal joints). While these joints are nearly never used in serial mechanisms, they are however omnipresent in hybrid mechanisms. The joints, often declared as being the key components of parallel and hybrid mechanism, are a central point in the development and research (75).

The development and evolution of 3 spherical joints is presented in the following sections. Their concept and design aimed for several qualities: **High stiffness** (higher than $300 \frac{N}{\mu m}$), **high mobility** (3 axes and $\pm 30°$ on each), **high precision** (low friction, precisely defined contact), **small size** (max. $\varnothing 50mm \times 100mm$), **low weight** and **low production costs** (lower than 1500 CHF).

---

[1] This way not modifying the mobility of the end-effector.
[2] Section 6.3 is concerned with the internal efforts and the design of segments.

6. MACHINE ELEMENTS

Because of the required, minimum strokes of ±30° flexure-based[1] joints were not considered. A gliding contact was chosen because of its large contact area (with respect to the overall joint size) at the interface of the 2 parts. In fact, as for all assemblies, the size and quality of the contact surface strongly determines the stiffness of the interface.

6.2.1 1st Evolution Spherical Joint

The first prototype principally aimed for a very high stiffness and was therefore mainly designed for having a big contact area. The following choices were made in order to maximize this contact:

Sphere-sphere interface: A high contact area can be achieved by choosing closely matching shapes. Two spherical parts with very similar diameters were chosen. The difference between the two diameters was machined as close as technically possible. The joint therefore consists in two main parts, having both a precisely machined spherical (a convex and a concave) surface. The contact between two spheres is a point-contact.

High Preload: By applying a preload between the 2 interfacing parts, the stiffness of the system can be enhanced, and the joint can be prevented from dislocating (for traction efforts exerted on the joint). The elastic deformation of the parts in contact, induced by the preload, increases the size of the contact area. Furthermore, preloading the joint allows

Limiting the amount of parts and interfaces: The joint consists of two principal parts, both of them have a machined spherical surface.

Figures 6.2 and 6.3 illustrate the 1st prototype.

Both *principal, spherical surfaces* ∅ *30mm* were machined with highest precaution in order to obtain the best possible surface finish and geometric precision. Both interfaces were analyzed and measured afterwards, the difference of both radius amounts to 0.02 mm, the sphericity is below 0.01mm. These surfaces determine the joints precision and act as mechanical reference.

The 2 parts in contact, the convex sphere and the concave spherical socket, are made

---

[1]The different joint technologies are presented in section 4.2. Simple flexure-based joints have a realistic application interval going from 0° to 10°. Special arrangements (Henein (36) and Pham (61), section 9.2) of multiple joints can increase the limits, however, they cannot be considered as simple joints.

6.2 Joints

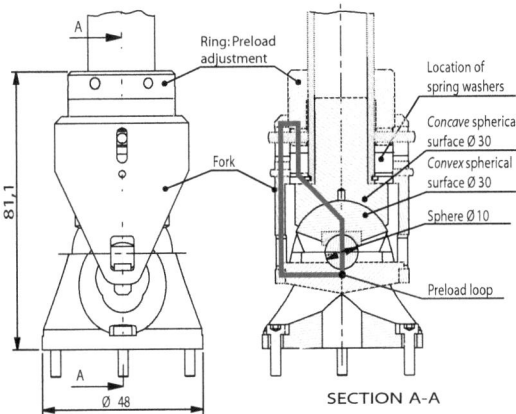

Figure 6.2: Drawing of the 1st evolution spherical joint. The mechanical loop transmitting the preload is indicated.

 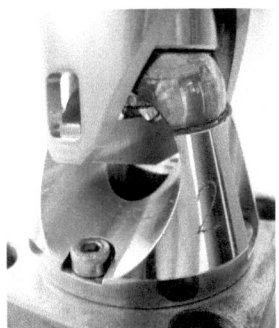

(a) Picture of the 1st evolution spherical joint. The spherical surfaces are machined in the basic components of the joint, therefore leading to relatively complicated parts

(b) Close-up on the contact region and the preloading system. The preload is applied on a smaller ball, by this reducing the friction by a factor 3 in the preloading system.

Figure 6.3: 1st evolution of spherical joint, designed in February 2005 and realized in April 2005.

6. MACHINE ELEMENTS

from stainless steel and a bronze alloy[1] respectively.
The *preload* on this interface is applied by means of a stack of elastic spring washers placed above the spherical socket. The force is transmitted through a fork on the backside of the convex sur, where a smaller ball (∅ 10mm) is placed. The smaller radius allows a great reduction of the friction in the preloading system (factor $\frac{30}{10} = 3$). The preload is adjustable by turning a ring on the upper side of the joint.

The *stiffness of the interface* was modelized using the theory of Hertz (22)(6) for the contact of two spheres (point contact, illustrated in figure 6.4). Considering one sphere fixed, the displacement of the center of the second sphere is defined by:

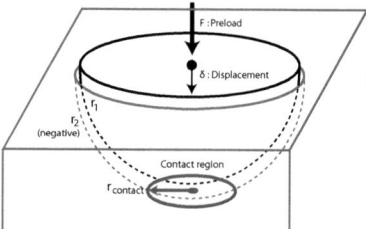

Figure 6.4: Illustration of the point contact and definition of parameters.

$$\delta = \left(\frac{9F^2}{4r_* E_*^2} \right)^{\frac{1}{3}} \quad (6.2)$$

where:

$$E_* = \frac{2E_1 E_2}{(1-\nu_1^2)E_2 + (1-\nu_2^2)E_1} \quad (6.3)$$

$$r_* = \frac{r_1 r_2}{(r_1 \pm r_2)} \quad (6.4)$$

E_i being Young's modulus and ν_i Poisson's ratio of the correspondent material, r_i the radius[2] of the spheres. Solving the equation for F and differentiating with respect to δ gives us the stiffness of the contact:

---

[1] Brush Wellman Toughmet3 CX Bronze alloy: $E_{bronze} = 129.9$ GPa, $\nu_{bronze} = 0.27$, measured friction coefficient on stainless steel $\mu = 0.05 - 0.1$ (lubricated with Molykote G Rapid Plus)

[2] In our case one of the radius (the bigger, r_2) has to be declared as negative value. This is because we consider one sphere as concave and enveloping the convex sphere. For convex-convex contact both radii are positive.

6.2 Joints

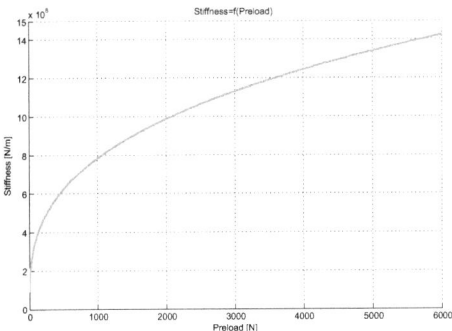

Figure 6.5: Stiffness of the contact between the 2 spheres depending on the applied preload. E_{steel}= 199 GPa, E_{bronze}=128.9 GPa, $\nu_{steel} = \nu_{bronze}$=0.27, $\varnothing_1 = 2r_1$=29.97mm, $\varnothing_2 = 2r_2$=-30.01mm.

$$k_{sphere-sphere} = \left(\frac{3}{2}\right)^{1/3} E_* \left(\frac{F^2}{E_*^2 r_*}\right)^{1/6} \sqrt{r_*} \qquad (6.5)$$

Furthermore, the radius of the contact region $r_{contact}$ and the maximum contact pressure σ_{max} are given by:

$$r_{contact} = \left(\frac{3Fr_*}{2E_*}\right)^{\frac{1}{3}} \qquad (6.6)$$

$$\sigma_{max} = \frac{3F}{2r_{contact}^2 \pi} \qquad (6.7)$$

Figure 6.5 shows the evolution of the stiffness with respect to the applied preload.

The stiffness of the spherical joint is not linear to the applied preload. The value of the preload F must be chosen higher than a possible traction force F_t (F and F_t are opposed). This, in order to prevent the dislocation of the joint and in order to keep a certain stiffness level.

The difference $\Delta r = r_1 - r_2$ between the 2 radii is greatly influencing the stiffness of the contact, figure 6.6 illustrates this. For small differences the size of the contact area strongly increases (stiffness proportional to $\Delta r^{-\frac{1}{3}}$). For very small difference of radius, the Hertzian model provides pessimistic results. Goodman (30) shows that the

6. MACHINE ELEMENTS

stiffness in such cases is slightly higher than predicted by Hertz. Considering that both radii r_1 and r_2 are very similar, and with a constant $\Delta r = r_1 - r_2$ (given by machining limitations), we can state $r_1 \approx r_2 \approx r$. By developing equation 6.5 and isolating r we notice that the stiffness of a point contact is proportional to $r^{\frac{2}{3}}$, which gives us an information about the impact of the joint's nominal size.

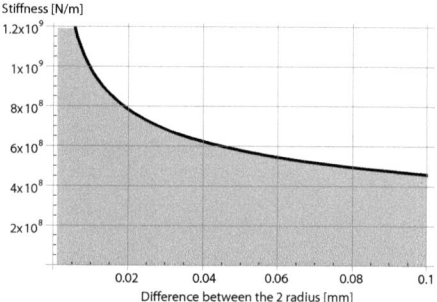

Figure 6.6: Stiffness of the contact depending on the difference of radius between the 2 spherical surfaces. E_{steel}= 199 GPa, E_{bronze}=128.9 GPa, $\nu_{steel} = \nu_{bronze}$=0.27, $\varnothing_1 =$ $2r_1$=29.97mm, Preload F=1000N.

For the above-mentioned reasons we measured the stiffness of the balljoint and compared the results with the prognosis coming from the Hertzian model. Figure 6.7 shows the measuring setup. Besides the interface, considering the way the force- and measuring loops are passing through the setup, the measures will include the stiffness of the mechanical parts of the spherical joint.

Since the Hertzian model does only take into account the stiffness of the contact $k_{sphere-sphere}$ we had to complete the model by introducing the elasticity of the parts k_{parts}. The stiffness of the main parts (convex sphere and the concave socket) were computed using FEM and estimated at $k_{parts} = 1538 \frac{N}{\mu m}$.

Since all these elements are arranged in series, we can state that the total compliance is equal to the sum of all individual compliances (16):

6.2 Joints

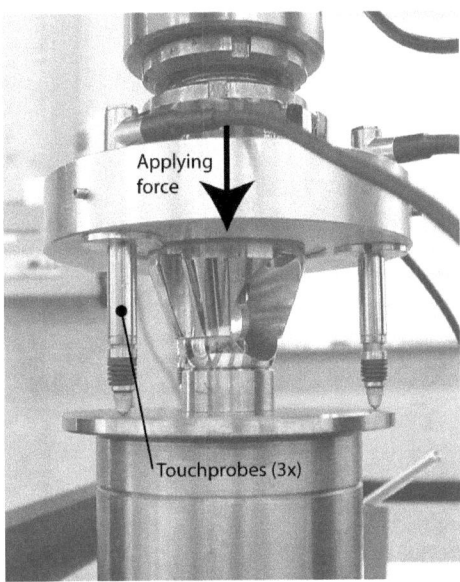

Figure 6.7: The setup used to measure the stiffness of the spherical joint. Three inductive touchprobes are monitoring the displacement that occurs between the 2 loaded parts. Using 3 touchprobes, and averaging their measures, filters out an eventual rotation taking place. The 2 interfaces between the spherical joint and the measuring machine were machined with precision and cleaned, guaranteeing the highest possible contact surface and stiffness. This partially avoids affecting the measure.

$$\frac{1}{k_{total}} = \sum \frac{1}{k_i} \qquad (6.8)$$

$$\frac{1}{k_{total}} = \frac{1}{k_{parts}} + \frac{1}{k_{sphere-sphere}} \qquad (6.9)$$

Giving us a term for the total stiffness of the system:

$$k_{total} = \frac{1}{\frac{1}{k_{parts}} + \frac{1}{k_{sphere-sphere}}} \qquad (6.10)$$

The comparison of our model and the measured values is illustrated in figure 6.8. The model and the measurement show a certain difference, especially for low preloads.

6. MACHINE ELEMENTS

Several reasons can be cited:

Right shift of the measurements: The measure exhibits a small right shift and slightly falsifies the comparison with the model. The setup probably needed a small load to get the parts correctly arranged at the beginning.

Non-perfect geometry of the joint interfaces: The machines joint interfaces do not present a perfect spherical geometry (geometrical errors, tool marks, surface roughness etc.) at micrometer scale. According to Spinnler (69), the contact occurs first on the asperities, whose amount increases with the applied preload, resulting in a linear preload-displacement behavior. This, for lower preloads, flattens the stiffness graph.

Interfaces between joint setup and measuring machine: These interfaces, although precise and clean, have a small influence on the measure (as they are located in the force- and measuring-loop).

Even with these slight differences, the models can be used to create predictions for future joint designs. The order of magnitude is correct, and this is what the designer is looking for in a first estimation of the characteristics.

Severe *wear tests* were conducted to verify the joint's behavior over a certain time period. For this reason the joint was constraint to move on a trajectory[1] for several days. The test has lasted for 1'300 hours and the joint has carried out 2'300'000 cycles. A visual check confirmed that, because of the wear, an enhancement of the contact surfaces happened. The measurement of the geometry showed that the diameter of the socket has adapted itself to the sphere (sphericity improved). The variation of the radius amounts to $5 \mu m$. According to our industrial partners, and according to the amount of executed cycles[2], such a small wear is absolutely tolerable.

Conclusion on the performances of the 1st evolution spherical joint

This 1st evolution of the spherical joint exhibits good mechanical properties (see table 6.1) but induces quite high production costs. In order to guarantee a high accuracy the contact surfaces implicate a precise and tedious machining. Besides the surfaces, the structural parts themselves are relatively complex and induce long machining times. The next evolution of joint should therefore present less production costs.

---

[1] The spherical socket was constrained on rotation cycles around the vertical axis with 5° inclination.

[2] The amount of executed cycles corresponds to 50 years of machining (non-stop) on a standard EDM machine. Anyway, during a such a long time period a machine tool is re-calibrated several times and the occurred wear is taken into account.

6.2 Joints

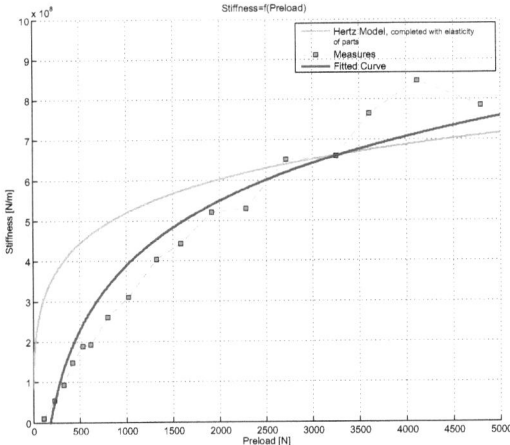

Figure 6.8: Stiffness of the 1st evolution spherical joint ($\varnothing$=30mm, Steel-Bronze interface, spherical shaped contact). The Hertzian model has been compensated with the stiffness of the mechanical parts of the joint.

Table 6.1: Characteristics of the 1st evolution spherical joint

Characteristic:	Value
Mass	0.588kg
Swivel angles	±30°
Stiffness@1000N Preload	$300\frac{N}{\mu m}$
Sphere $\varnothing$	30mm
Required space	$\varnothing$48mm x 81mm
Estimated prod. costs	800-900.- CHF

Another thing that was noticed, while characterizing this articulation, was the relative influence of the interface's stiffness with respect to the total stiffness. The interface should not be neglected but its stiffness should be balanced with the other elements. Figure 6.9 graphically compares the single and total stiffness of a typical arrangement of elements.

6. MACHINE ELEMENTS

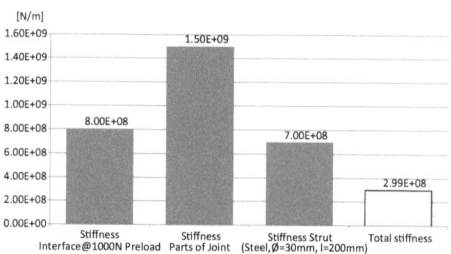

Figure 6.9: Comparative study of the stiffness of different elements in a kinematic chain. The chain is composed of a spherical joint and a strut, a typical sequence of elements in a parallel/hybrid mechanism. The stiffness of a serial arrangement of elements can be computed according to equation 6.9.

This fact relativizes the importance of having an extremely stiff interface, thus, a minimum value should be obtained since, in a system of serially-arranged compliances, the total stiffness will always be lower than its weakest element.

6.2.2   2nd Evolution Spherical Joint

The 2nd evolution, which is illustrated in figure 6.10, presents two crucial differences compared to the 1st evolution:

Independent ball: The convex spherical surface is not machined anymore. A sphere, initially intended as bearing ball, is now the central convex element of the joint. These elements are among the most precise mechanical parts available, and, since they are produced in huge amounts, are relatively inexpensive. Because of their given accuracy they won't be machined in any way, forcing the joint to possess 2 interfaces.

Interface: The sockets are shaped as cones, which simplifies a lot their machining. The resulting contact type between ball and cone is a line contact, a new model will therefore be created.

Simple and effective design were the keywords for the design of this 2nd version. The whole joint consists of 3 main parts: Two cones being the output and the input of the joint and the central sphere. The preloading system only consists of a spring. In order to obtain a sufficient preloading force, and, in order to bend according to the movement of the joint ($\pm 30°$), the spring becomes quite long.

6.2 Joints

Figure 6.10: 2nd evolution of the spherical joint.

The interface is modeled using the formulas developed by Roark (77). The model used describes the displacement of a cylinder which is pressed on a plane. This model has been adapted for the case of a *sphere-cone* whose definitions are shown in figure 6.11. In order to adapt the model to our case, we unwind the sphere and the cone to obtain a cylindrical surface on a plane. The length of the contact line becomes:

$$L_c = \frac{\pi}{2} d \cos(\frac{\alpha}{2}) \tag{6.11}$$

with a projected force of:

$$W = \frac{F}{2\sin(\frac{\alpha}{2})} \tag{6.12}$$

According to Roark (77) the displacement of the center, already projected back on the vertical direction, becomes:

$$\delta = \frac{C_E W}{\pi L_c \sin\left(\frac{\alpha}{2}\right)} \left(\frac{1}{3} + \ln\left(\frac{2d}{b_c}\right)\right) \tag{6.13}$$

where:

$$b_c = 1.5 \sqrt{\frac{W d C_E}{L_c}} \tag{6.14}$$

$$C_E = \frac{1 - \nu_1^2}{E_1} + \frac{1 - \nu_2^2}{E_2} \tag{6.15}$$

6. MACHINE ELEMENTS

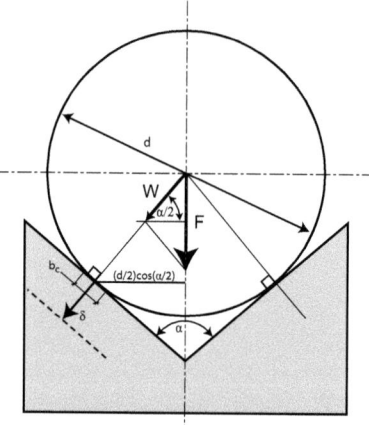

Figure 6.11: Definition of the parameters for the modeling of the contact.

b_c being the width of the contact line, E_i the Young modules of the two facing materials, α the aperture angle[1] of the cone and ν_i the corresponding Poisson coefficients.

By differentiating δ according to F we obtain the stiffness of the interface:

$$k_{\text{sphere-cone}} = \frac{12\pi L_c \sin^2\left(\frac{\alpha}{2}\right)}{C_E \left(6\ln\left(\frac{2d}{1.6}\left(\sqrt{\frac{C_E dF}{2L_c \sin\left(\frac{\alpha}{2}\right)}}\right)^{-1}\right) - 1 \right)} \qquad (6.16)$$

The total stiffness of the complete joint becomes:

$$k_{total} = \frac{1}{\frac{1}{k_{cone1}} + \frac{1}{k_{sphere-cone}} + \frac{1}{k_{sphere}} + \frac{1}{k_{sphere-cone}} + \frac{1}{k_{cone2}}} \qquad (6.17)$$

The computed stiffness as well as the measurements are presented in the next subsection.

---

[1] An optimal angle, considering the stiffness, can be found around $\alpha = 100°$. This is due to the projection of the force $W(\alpha)$, the projection of the displacement $\delta(\alpha)$ and the length of the contact line $L_c(\alpha)$, all of them depending on the aperture angle.

6.2 Joints

Enhancing the stiffness and quality of the interface by stamping

In order to enhance the contact surface and thereby the stiffness of the interface, a special process was developed. It consists in *stamping the interface of the cone by applying a force, high enough to plastically deform the interface*. The force during the stamping is transmitted through a hard-metal tungsten-carbide sphere (WC) which avoids the sphere from being deformed plastically, and guarantees a spherical shape in the cone. An example of two stamped cones is illustrated in figure 6.12. The resulting surface on the cone has an excellent surface finish, similar to a polished surface.

Figure 6.12: Picture of two cones (bronze-alloy and steel, for a 15mm ball) after a stamping of 50kN on a hydraulic press. The stamped surface is visible as a ring on the cones.

The stiffness measured for different stamping forces, and with respect to the preload of the joint, is represented in figure 6.13.

The stamping process enhanced the total stiffness of the joint by 32% (if we look at the stiffness for a 1000N preload). Considering that the stiffness of the structural parts did not change during this operation, we can compute (using equation 6.9) an increase of a single interface stiffness: *The interface stiffness increased by 200%*.

Another considerable augmentation of the overall stiffness can be obtained by placing – instead of the usual steel ball – a tungsten-carbide ball. This exchange has 2 major impacts:

Enhanced structural stiffness: Since the Young modulus is 2-3 times ($E_{WC} =$

6. MACHINE ELEMENTS

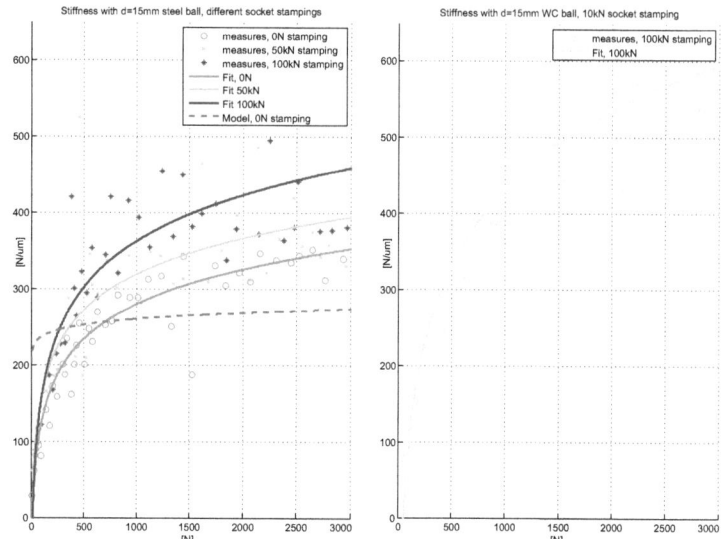

(a) Stiffness of the joint as a function of the preload, and for different stamping forces. Both, the cones and the ball are made from steel.

(b) Stiffness of the joint with a tungsten-carbide ball.

Figure 6.13: Stiffness of the 2nd evolution spherical joint. Both cones are still made from steel.

$640 GPa$) higher than steel, the elasticity of the sphere augments by the same amount.

Enhanced interface stiffness: The stiffness of the interface depends on the Young modulus of both adjacent materials. A higher modulus greatly enhances the stiffness of the interface (equation 6.15).

Here again, as for the 1st evolution, the physical model shows some differences especially for low preloads. In general however, it predicts correctly the order of magnitude of the stiffness.

6.2 Joints

Conclusion on the performances of the 2nd evolution spherical joint

The second evolution spherical joint possesses very similar mechanical properties as the 1st version. Its performance is summarized in table 6.2. However, because of its mechanical concept, it greatly reduces the complexity of the mechanics.

A very effective and simple process to enhance the interface stiffness was developed. By stamping the interface, and by using a tungsten-carbide (WC) ball, we obtain slightly higher stiffness results than the 1st evolution spherical joint, even with a 2 times smaller sphere diameter.

Table 6.2: Characteristics of the 2nd evolution spherical joint, $\varnothing = 15mm$, with tungsten-carbide (WC) ball and for a stamping force of 100kN.

Characteristic:	Value
Mass	1.27kg
Swivel angles	$\pm 30°$
Stiffness@1000N Preload	$430 \frac{N}{\mu m}$
Sphere $\varnothing$	15mm
Required space	$\varnothing$36mmx120mm
Estimated prod. costs [CHF]	n.a.[a]

[a]The production costs of this version were not investigated because an enhanced version was planned anyways.

Some few drawbacks of the preloading system – the spring – still motivated us to reconsider certain mechanical details and develop a 3rd evolution. The restoring force can create internal constraints (bending moments, see figure 6.14) in the segments of the mechanism and influence its precision. Furthermore, the size of the spring considerably increases the height of the spherical joint.

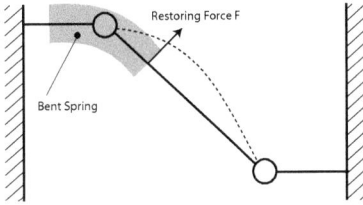

Figure 6.14: Preloading spring induces a bending moment in a kinematic chain.

6. MACHINE ELEMENTS

6.2.3 3rd Evolution Spherical/Universal Joint

The 3rd evolution only reconsiders the preloading system of the joint. The interfaces, their processing and the spheres are kept the same as for the previous evolution.

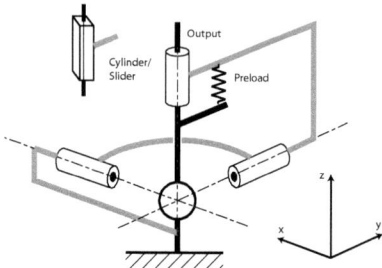

Figure 6.15: Simplified kinematics of the 3rd evolution of the spherical/universal joint. Depending on the working mode the cylinder joint can easily be converted to a slider joint.

The preloading system was designed as a second "joint" around the central sphere (see figure 6.15). It is constituted of 2 simple pivots which are arranged as a universal joint in the horizontal plane, around the sphere. A third joint is arranged vertically, in a collinear way to the output of the joint. The preload on the central sphere is created by applying a force between the outer "joint" and the central sphere. The force is created by a stack of spring washers but can theoretically be generated by any elastic element. The final joint is represented in figure 6.16.

A coupling, placed on the upper part of the joint, allows to lock the rotation of the vertical cylinder joint. This way, instead of allowing a rotation around the vertical axis (torsion), the whole joint behaves like a universal joint, transmitting all torsion loads through the outer "joint". The coupling of the central ball joint and the outer preloading system is achieved by means of a membrane, see figure 6.17. The membrane is designed for having a high stiffness in its plane (transmitting torque) and a high compliance in the other directions, preventing the whole joint to be highly overconstrained. The mode of operation, spherical joint or universal joint, can therefore be chosen, and set up, by removing just 2 screws.

The whole joint can be roughly divided in 2 parts:

The joint core which is constituted as for the 2nd evolution, consisting of 2 cones

6.2 Joints

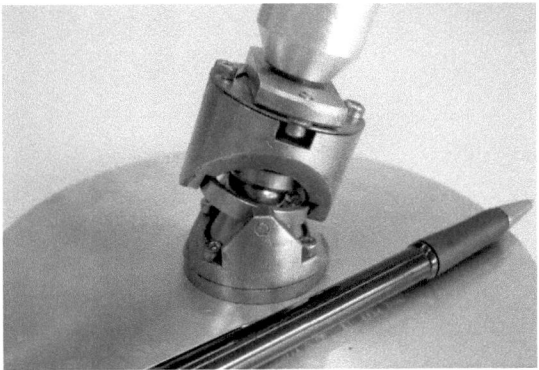

Figure 6.16: 3rd evolution of the spherical joint. First concepts of this type of joint were proposed in June 2004 (45).

and the sphere. The components and interfaces of this part define the stiffness and the precision of the joint, and would bear all dynamic efforts generated by the moving machine.

The outer joint applies the preload and transmits the torsion, in case of setting up the articulation as universal joint. This part of the joint does not contribute to the precision of the joint, its design only considers constraints and loads.

Conclusion on the performances of the 3rd evolution spherical joint

Table 6.3 below resumes the characteristics of the 3rd evolution spherical joint.

Besides having very satisfying mechanical properties this joint provides 2 modes of operation without changing the mechanics. A same compact design can be used to create a spherical joint or a universal joint. The torsion stiffness was estimated around 6'350 $\frac{Nm}{rad}$ which corresponds, as comparison, to the torsion stiffness of a ballscrew with $\varnothing = 20mm$ and $l = 200mm$. This one-design joint fulfils most requirements, with respect to mobilities and mechanical characteristics, of parallel/hybrid mechanisms.

6. MACHINE ELEMENTS

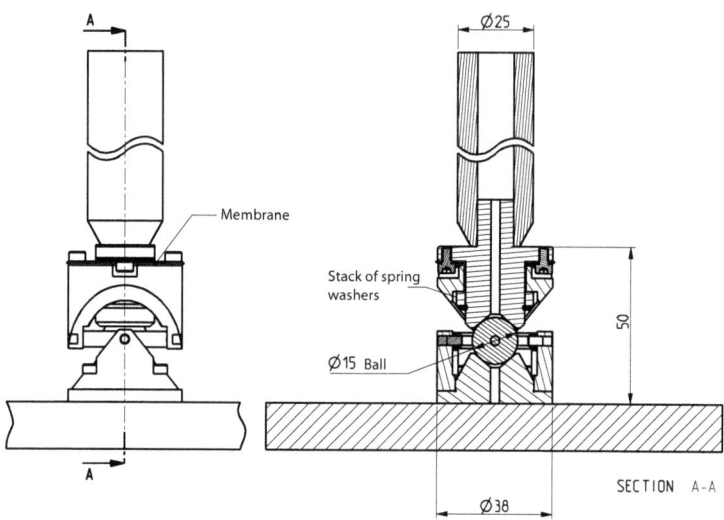

Figure 6.17: Section through the 3rd evolution spherical joint. Note: In case of an industrialized design the upper tube and the adjacent part of the spherical joint should be designed as 1 part only.

Table 6.3: Characteristics of the 3rd evolution spherical joint, $\varnothing = 15mm$, with tungsten-carbide (WC) ball and for a stamping force of 100kN.

Characteristic:	Value
Mass]	0.225kg
Swivel angles	±30°
Stiffness@1000N Preload	$430\frac{N}{\mu m}$
Torsion Stiffness	$6'350\frac{Nm}{rad}$
Sphere ⌀	15mm
Required space	⌀38mmx50mm
Estimated prod. costs	500-600.- CHF
Special feature	2 or 3 DOF, easily interchangeable

6.2.4 Conclusion of the section

If we now compare the last version of our spherical joint, the 3rd evolution, to the commercially available INA GLK balljoint (small version, see table 4.1) we can notice a 53% higher stiffness ($430\frac{N}{\mu m}$ to $280\frac{N}{\mu m}$), a 79% smaller volume ($57cm^3$ to $270cm^3$) and a 90% lower mass (0.225kg to 2.3kg) of the joint. Both joints having nearly the same swivel angles. Furthermore, no commercially available joint proposes a system to switch between 2 modes of operation.

The development of all prototypes has led to a well performing spherical/universal joint of small size and reasonable complexity. A lot of importance was been given to simple and common production techniques, as for example the stamping, in order to design a industrially viable joint.
All different contact types were modelized mathematically in order to help the designer estimate the final stiffness of his joint design. The models are precise enough to predict the order of magnitude. While modeling the interface we could observe that, for the interface stiffness that we can reach with the stamping technique, the structural stiffness of the mechanical parts becomes crucial too.
The 3rd evolution spherical joint has been integrated in the design of two prototypes, the Stewart3 and the Omikron5. These machines are presented in sections 9.1 and 9.3 respectively.

Figure 6.18: All 3 evolutions of the spherical joint. The 4th version (on the right) is the industrialized and simplified spherical joint.

6. MACHINE ELEMENTS

6.3 Segments/Rigid Bodies

The structural elements, often called segments, interconnect the joints of a mechanisms. They are exposed to dynamic loads generated by the machine's inertia or by static loads induced by the machining process or gravity, therefore, they are subject to internal efforts. As for the joints, their stiffness is very determining for the overall stiffness of the machine. This chapter will give a short overview on the most represented load cases and on efficient possibilities to enhance the stiffness of single segments. Furthermore, we will enumerate design principles which, on a higher level, can lead to a more effective global machine design.

Since this subject is very complex, and depending on the shapes of the considered parts, we won't give any mathematical solutions for the deformations[1] of parts, but instead, point out the crucial dimensions which highly influence the amount of deformation (and therefore the stiffness). The goal of this section is to *sensitize the machine designer to the load cases and the possibilities he has to enhance the stiffness of the parts*.

For this, we will consider an elastic element of rectangular shape (with a Young modulus E and a sliding coefficient G) which defines a geometric parameter L, therefore fixed in length, and whose main axis is parallel to L.

Traction/compression (figure 6.19, left side) The constraints generated by this load case are aligned to the force F and are defined by $\sigma = \frac{F}{A} = \frac{F}{BH}$. The stiffness of an element which is exposed to traction/compression is proportional to the surface $A = BH$ and inversely proportional to L.

Torsion (figure 6.19, right side) The torsion stiffness of an element which is exposed to a torsion torque is proportional to the polar moment of inertia[2] given by (for a rectangle) $\frac{BH}{12}(B^2 + H^2)$ and inversely proportional to L. The polar moment of inertia contains exponents equal to 3 which shows the huge gain obtained if B and H can be increased.

Simple flexion (figure 6.20) The effect of the flexion moment can be reduced by increasing the moment of inertia $\frac{BH^3}{12}$. This term contains exponents equal to 3 for parameter H.

---

[1] There exist a multitude of deformation cases, depending on the amount and type of attachment points and loads. The most represented can be found in technical literature (e.g. (28))
[2] Of the surface.

6.3 Segments/Rigid Bodies

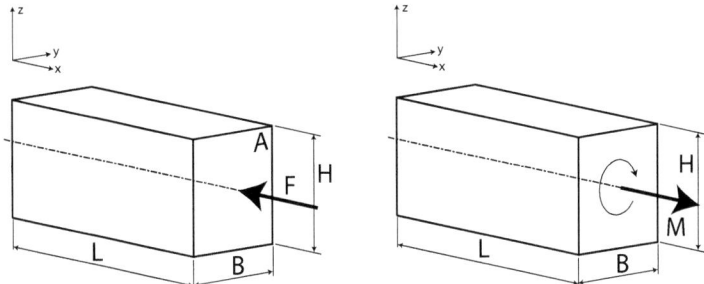

(a) Traction/compression load case. A force F, aligned to the main axis, acts on the surface A.

(b) Torsion load case. A moment M, aligned to the main axis, acts on the part.

Figure 6.19: Traction/compression and torsion load cases.

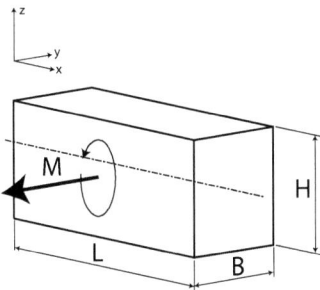

Figure 6.20: Simple flexion load case. A moment M, perpendicular to the main axis and created by a localized force or pressure distribution, acts on the part.

Different moments of inertia, corresponding to different section shapes, can be found in the literature (29)(28).

According to the **theorem of superposition**, for linear and isotropic materials, and for composed load cases, *the total deformation is equal to the sum of all single deformations considered individually*. Composed load cases can therefore be divided in basic cases and their stiffness can be investigated separately.

6. MACHINE ELEMENTS

The simple cases mentioned in this section allow the designer to point out the weak points in the design of a segment, and effectively improve it by modifying the most influencing parameters. After a first evaluation of a mechanism, the designer can then make use of a FEM system (Finite Element Method) in order to perfect his design. To determine the load case that has to be applied to a segment a static (applying efforts on end-effector) and dynamic (considering dynamic efforts due to inertia) analysis of the mechanism has to be carried out.

On a higher level, when considering several segments and their interaction, a few design principles regarding stiffness issues can be mentioned[1]. All of these principles have one common, crucial point which is about **avoiding to create additional internal efforts**[2]:

Directing efforts in traction/compression: As often as possible trying to direct the internal efforts of the mechanism in traction/compression mode, which is the most effective and simple way to increase the stiffness. Avoiding bending and torsion as much as possible. An example is presented in figure 6.21.

Aligning efforts directly on their supports: Aligning efforts with the sliders, ball-screws, successive segments etc. in all possible directions. If no perfect alignment is possible, reducing the size of the overhangs as much as possible. An example is presented in figure 6.22.

Keeping force loops as short as possible: Shortening the total length on which efforts are applied, by this reducing the deformation or even avoiding additional efforts. An example is presented in figure 6.23.

---
[1] This listing does not pretend to be exhaustive, but it points out the most important tendencies.
[2] And thereby avoid creating additional compliances.

6.3 Segments/Rigid Bodies

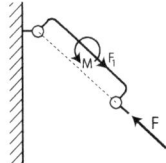 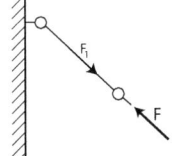

(a) The segment is not aligned with the load, an additional flexion moment M is created.

(b) The segment is aligned in an optimal way.

Figure 6.21: Directing efforts in traction/compression. F represents an arbitrary load whereas F_1 represents the internal effort.

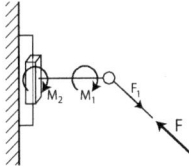 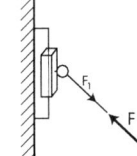

(a) The segment, which offsets the spherical joint from the slider, induces an additional flexion moment on itself (M_1) and on the slider (M_2).

(b) Avoiding the overhang directly aligns the effort on its support. Less (or even none) flexion is generated.

Figure 6.22: Aligning efforts directly on their supports.

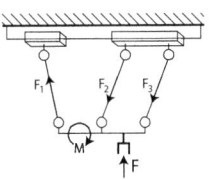 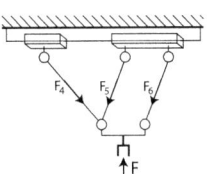

(a) The distance between the 2 outer-left joints on the mechanism's output creates an additional flexion moment, although this length is not vital for the mechanism.

(b) The distance has been reduces to its minimum by aligning the 2 joints on a single point.

Figure 6.23: Keeping force loops as short as possible.

6. MACHINE ELEMENTS

6.4 Summary and Conclusion of the Chapter

This chapter is concerned with the mechanical elements of a mechanism: The kinematic chains, the joints and the segments. Kinematic chains, which are treated in section 6.1, guide the mechanism by providing movement constraints, and on the other hand provide mobilities to actuate the end-effector. The concept of non-constraining kinematic chains is introduced. This concept allows the mechanical decoupling of actuation and guiding and, allows a modification of serial kinematics in order to obtain parallel kinematics.
In section 6.2 new concepts, models, and a new manufacturing process for high-mobility joints are presented. The last version of joint presents good performance compared to commercially available joints. Furthermore, the same joint design allows two working modes, the 3-DOF spherical joint mode or the 2-DOF universal joint mode. Both types of joints are often represented in hybrid kinematics and can be achieved with a single design.
Section 6.3 treats the rigid bodies which interconnect the joints. The different load cases of segments in hybrid kinematics are illustrated and the crucial dimensions to increase the stiffness are isolated. On a higher level, considering the interaction of multiple segments, design principles are provided. All principles have a common goal namely to avoid the creation of additional internal efforts in the mechanism.
The whole chapter provides a base for a purposeful mechanical design of hybrid and parallel kinematics. It concepts as well as the proposed joint designs were used for the development of 3 machines. They are presented in chapter 9.

7
Aspects of Industrialization

This chapter compiles a series of important design steps and choices a designer will have to make for an industrialized design. It starts with very essential questions, like the distribution of all mobilities on both hands of the machine, and ends with concepts around the final characterization measurements of the prototype.

7.1 Distribution of the DOF on the kinematics

The distribution of the total DOF on the 2 hands of a mechanism highly influences the global behavior of the machine. From a mechanical point of view it is interesting to distribute the DOF (76). The total mobile mass can be distributed on 2 mechanisms and the individual modules have less dynamic efforts to bear. Compared to a serial arrangement of these modules and their masses, the eigenfrequency would be considerably higher. Furthermore, the splitting of the total mobility can provide simpler kinematics and by this reduce the complexity, the costs and increase the reliability of the modules.

However, depending on the application, a distribution is not always possible as there exist applications/processes which impose it. The machining of very large and heavy pieces for example (machining large stamping dies, laser cutting of small parts from large, blank sheet) partially or completely inhibit the left hand from moving. Other limitations can be given, for example when adding movement capabilities to an existing machine by integrating additional axes. There might be no choice because the existing machine cannot be modified, or, because there is simply no space on one hand.

This section is therefore intended for design studies which allow this investigation. Several cases will be discussed in order to illustrate the possibilities and consequences of different distribution strategies.

7. ASPECTS OF INDUSTRIALIZATION

7.1.1 Decoupling the DOF

A mechanical decoupling of DOF can be achieved by splitting the total DOF on the two hands. The 2 modules, placed each on a hand, can obtain *dedicated designs* according to the type of DOF they execute. A typical, but not the only, strategy is to decouple the translations from the rotations by placing them on separate hands. A good example is the hybrid-kinematic Verne machine tool (figure 4.11 in chapter 4). A decoupling can also be beneficial if, for some sub-operations of the process, only one hand has to be moved, leaving the other static.

7.1.2 Identic modules

If both, left and right hand, have to execute the same amount and type of DOF a common design can be considered. Figure 7.1 shows a machine concept which is constituted of 2 identic modules. In order to generate different DOF, one of the modules has been rotated by 90°. Concerning the illustrated machine, and in order to allows the creation of 2 identical modules, a redundant translation was created[1]. The machine can execute movements according to 5 axes but possesses 6. The designer might prefer adding a redundant, unnecessary axis, but on the other hand allow an absolutely identic design for the 2 modules, reducing by this the manufacturing and design costs.

7.1.3 Different, special axis characteristics

If certain axes require strongly differing characteristics (e.g. high acceleration, high stiffness, different movement amplitudes etc.) or a different technology (different joint technology or drive technology) than the remaining axes they can be separated/isolated by placing it on the opposite hand. The drilling of small holes in μEDM requires a very fast and repeated insertion-movement on one axis. The insertion movement could be created by means of direct drive linear motor placed on one hand, and where the remaining axes, placed on the opposite hand, could be slower and using a more conventional technology. Another example could be related to tool-changing, where one axis, with an unusual high stroke, is performing the tool-changing.

The distribution of the mobilities can allow the regrouping of axes with similar characteristics and the separation of technologies.

---

[1]Fortunately, in this case it was beneficial, since a higher stroke was needed in this direction anyways.

7.1 Distribution of the DOF on the kinematics

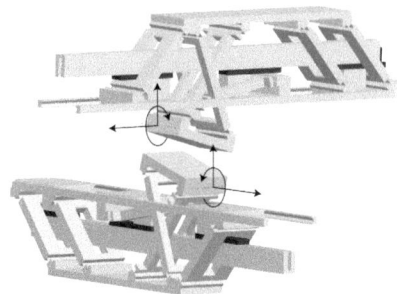
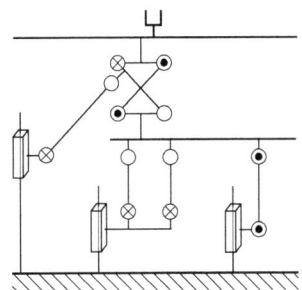

(a) A 5-axes concept, developed by Regamey, using 2 identical modules on the left and right hand. Each module generates the 2 translations is its main plane and a rotation perpendicular to the plane. When rotating one module by 90°, with respect to the other, we obtain a 5 axes machine (with a redundant translation).

(b) Isostatic kinematic scheme of one hand. Mechanism illustrated on the left however uses only pivots.

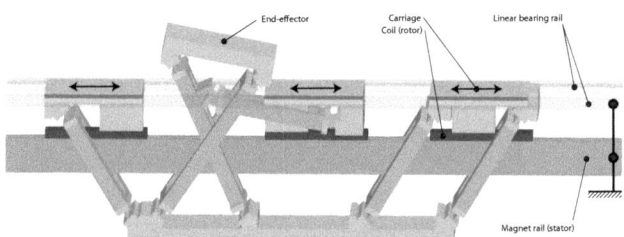

(c) A single, planar module uses a common linear bearing rail system and a common magnet rail for the 3 axes of the module.

Figure 7.1: A 5-axes concept using twice the same module. The module is designed for flexures and allows high rotation amplitudes around ±15° (without exceeding ±7.5° on any pivot). (a) presents the combination of 2 modules, (b) the kinematics and (c) a single, detailed module.

7.1.4 Distribution possibilities for 3-6 axes mechanisms

The following tables (7.3 - 7.11) illustrate all the possibilities of distributing 3-6 DOF on the 2 hands of the mechanism. The tables are classified according to the total amount of DOF. Existing parallel/hybrid machines are listed in their corresponding

7. ASPECTS OF INDUSTRIALIZATION

fields. No differentiation is made between symmetric distributions, the right and left hand can be inverted in any case. According to the project requirements, especially the imposed amount of mobilities, a designer can be inspired by all possibilities provided in the corresponding table.

Notations: A total amount of n DOF will be distributed on 2 hands and written as $i\mathbf{T}j\mathbf{R}$, which defines the needed type and amount of mobilities. "**T**" stands for the *Translation* and "**R**" for *Rotation*. Logically, $(i+j)$ must be equal to n.

For each axis-distribution $i\mathbf{T}j\mathbf{R}$, and without considering symmetric distributions, there exist $N_{i\mathrm{T}j\mathrm{R}}$ possibilities[1], defined as:

$$N_{i\mathrm{T}j\mathrm{R}} = \frac{1}{2}(i+1)(j+1) \tag{7.1}$$

Table 7.1: 3 axes: 3 translations: **3T**

Right Hand	T	3	2
	R	0	0
Left Hand	T	0	1
	R	0	0
Mechanisms		Delta	

Table 7.2: 3 axes: **2T1R**

T	2	2	1
R	1	0	1
T	0	0	1
R	0	1	0

Table 7.3: 3 axes: **1T2R**

T	1	1	1
R	2	1	0
T	0	0	0
R	0	1	2
		Orion, Dunlop (figure 9.10, section 9.2)	

Table 7.4: 3 axes: **3R**

T	0	0
R	3	2
T	0	0
R	0	1
	Agile Eye (31)	

---

[1] Some cases require to round up the result to the next integer value.

7.1 Distribution of the DOF on the kinematics

Table 7.5: 4 axes: 3 translations and 1 rotation: **3T1R**

Right Hand	T	3	3	2	2
	R	1	0	1	0
Left Hand	T	0	0	1	1
	R	0	1	0	1
Mechanisms		Delta4, Mitsubishi RP			

Table 7.6: 4 axes: **2T2R**

T	2	2	2	1	1
R	2	1	0	2	1
T	0	0	0	1	1
R	0	1	2	0	1

Table 7.7: 4 axes: **1T3R**

T	1	1	1	1
R	3	2	1	0
T	0	0	0	0
R	0	1	2	3

Table 7.8: 5 axes: 3 translations and 2 rotations: **3T2R**

Right Hand	T	3	3	3	2	2	2
	R	2	1	0	2	1	0
Left Hand	T	0	0	0	1	1	1
	R	0	1	2	0	1	2
Mechanisms		Exechon, Space5H, Pegasus, Alpha5, Tricept, Omikron5	Ecospeed, Dumbo	Hita STT	Verne, Vision	Genius 500, Trijoint 900H	

Table 7.9: 5 axes: 2 translations and 3 rotations: **2T3R**

T	2	2	2	2	1	1
R	3	2	1	0	3	2
T	0	0	0	0	1	1
R	0	1	2	3	0	1

7. ASPECTS OF INDUSTRIALIZATION

Table 7.10: 6 axes: 3 translations and 3 rotations: **3T3R**

Right Hand	T	3	3	3	3	2	2	2	2
	R	3	2	1	0	3	2	1	0
Left Hand	T	0	0	0	0	1	1	1	1
	R	0	1	2	3	0	1	2	3
Mechanisms		Stewart							

Table 7.11: **Special redundant case**, 6 axes: 4 translations and 2 rotations: **4T2R**

T	4	4	4	3	3	3	2	2
R	2	1	0	2	1	0	2	1
T	0	0	0	1	1	1	2	2
R	0	1	2	0	1	2	0	1
						MinAngle, Stewart3 + translator		Mechanism on figure 7.1

7.2 Geometric Modeling

The geometric modeling is about the creation of the direct (DK) and inverse (IK) kinematic models. The definition of these models can be found in section 3.2.4. This modeling is necessary for the future control of the robot, but it can also be very useful for certain optimization tasks during the mechanical design. It consists in finding the geometric relationships between the operational coordinates X and the articular coordinates Q.

In a lot of cases these models can be solved analytically, but unfortunately some kinematics require a numerical approach:

Mechanisms with pure mobilities possess kinematics that *totally decouple the actuated operational coordinates from the constrained coordinates*. Pure translators for example (e.g. the Delta) do not modify[1] the remaining rotational DOF when moving. Their kinematics are normally modeled analytically.

Mechanisms with non-pure mobilities possess kinematics that *affect more than just the actuated operational coordinates during movement*, they have so-called *parasitic movements*. The MinAngle's left hand (see section 9.2) for example has a very limited – but existing – translational movement when executing a rotation. If there exist a analytical solution, it can be extremely difficult to obtain, a numerical approach is therefore preferred.

The creation of analytical geometric models is well-known and already well documented, Merlet (48) proposes the solutions for a lot of different kinematics. But as most prototype developed during this thesis possess non-pure mobilities, this chapter will focus on the numerical solution of the inverse and direct kinematic problems. The numerical solving of the DK and IK is a straight-forward and general approach. The main difference to the analytical solving lies in the handling of the non-linear equations. Whereas the analytic solving (of mostly high-order polynomials) needs a lot of mathematical finesse to isolate the desired coordinate, the numerical solving uses a iterative algorithm to approach the solution[2]. From a industrialization point of view, a numerical approach can be implemented and handled faster.

---

[1] Theoretically, when considering the mechanism's geometry as being perfect.
[2] A closed-form analytic model is however more valuable. It is greatly useful to optimize the geometric parameters for example, or, to detect singularities.

7. ASPECTS OF INDUSTRIALIZATION

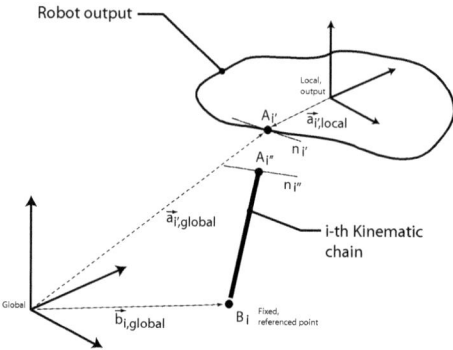

Figure 7.2: Definition of reference systems and points for the mathematical modeling. The kinematic chain is represented in a simplified manner, it can take very complex shapes.

Figure 7.2 sets the background definitions for the following explanation.

The creation and solving of a numerical model has 2 distinct parts: the creation of the – mostly non-linear – equations and system of equations, and the iterative algorithm solving the system.
The equation systems of the 3 developed prototypes (see chapter 9) are carried out in appendix C.

7.2.1 Creating the Equation system

Notations: Scalars are indicated as normally typeset letters (a), **vectors** as bold letters $(\mathbf{A})$, **matrices** as bold letters with their dimensions as indices $(\mathbf{A}_{nxn})$.

Creating the equation systems is about finding *2 different ways of describing a same variable in the kinematics*. This object can be either a point, an axis, or any other entity. The difference (subtraction) between the 2 distinct approaches creates an equation system whose solution will be found by an iterative algorithm.

We first define the actuated operational coordinates $\mathbf{X}$ and the remaining, parasitic coordinates $\mathbf{P}$, then generate the general 3x3 rotation matrices $\mathbf{R}_{3x3}$ (which contains the rotative operational coordinates) and the general 3x1 translation vector $\mathbf{T}_{3x1}$ (which contains the translational operational coordinates). Furthermore, we fix a *local refer-*

7.2 Geometric Modeling

ence frame to the end-effector.
Then, we apply the transformations to a point on the robot's output using homogenous matrices. This gives us a *global* description of the output, and of the point, introducing all operational coordinates.

$$\mathbf{A}_{I'}(\mathbf{X}, \mathbf{P}) = \begin{pmatrix} a_{ix} \\ a_{iy} \\ a_{iz} \\ 1 \end{pmatrix}_{global} = \begin{pmatrix} \mathbf{R}_{3x3} & \mathbf{T}_{3x1} \\ 0 & 1 \end{pmatrix} \cdot \begin{pmatrix} a_{ix} \\ a_{iy} \\ a_{iz} \\ 1 \end{pmatrix}_{local} \quad (7.2)$$

Where i is the index indicating the i-th kinematic chain ($i = 1..m$). Which gives us the 1st description of the point $\mathbf{A}_{I'}$. The point $\mathbf{B}$ is a known, fixed starting point of the kinematic chain.

Starting from the point $\mathbf{B}$ we describe the kinematic chain considering all constraints (e.g. perpendicularity constraints of pivots, lengths of segments) given by the different joints. The articular coordinates $\mathbf{Q}$ are introduced in the description of the kinematic chain. Now we have a second description for the point $\mathbf{A}_{I''}$ and can state the equation:

$$|\mathbf{A}_{I'}\mathbf{A}_{I''}| = 0 \quad (7.3)$$

Depending on the complexity of the kinematics this step may have introduced additional unknowns $\mathbf{W}$, like passive joint angles. This step depends on the kinematics and will be detailed in appendix C, applied to the developed prototypes.
If the $\mathbf{A}_{I'}$ was located in a pivot we also need to match the joint axes $\mathbf{n}_{I'}$ and $\mathbf{n}_{I''}$ which gives us a second equation for this kinematic chain. This is done using the scalar product of the 2 vectors describing the axis direction, again, approaching from 2 different ways:

$$\mathbf{n}_{I'} \cdot \mathbf{n}_{I''} - 1 = 0 \quad (7.4)$$

In general, a kinematic chain containing a pivot is constraining more strongly the output than a kinematic chain having only spherical joints. This results in a logically lower amount of kinematic chains, less equations where 2 points can be matched (type: $\mathbf{A}_{I'} = \mathbf{A}_{I''}$) but adds equations where joint axes have to match[1] (type: $\mathbf{n}_{I'} \cdot \mathbf{n}_{I''} = 1$). The MinAngle (section 9.2) perfectly shows that.

The system of equations $\mathbf{F}$ can now be generated.

---

[1] A Stewart platform for example has 6 kinematic chains without any pivot. Its kinematics are modeled using 6 equations of type $\mathbf{A}_{I'} = \mathbf{A}_{I''}$ (distance constraint), since no alignment of axes is necessary.

7. ASPECTS OF INDUSTRIALIZATION

$$\mathbf{F} = \begin{pmatrix} f_1 \\ f_2 \\ . \\ . \\ f_n \end{pmatrix} = \begin{pmatrix} |\mathbf{A_{1'}(X,P)A_{1''}(W,Q)}| \\ . \\ |\mathbf{A_{m'}(X,P)A_{m''}(W,Q)}| \\ \mathbf{n_{1'} \cdot n_{1''}} - 1 \\ . \\ \mathbf{n_{m'} \cdot n_{m''}} - 1 \end{pmatrix} = 0 \quad (7.5)$$

It contains n equations, n being the amount of unknowns. Translated into equations, an isostatic systems has the same amount of unknowns and equations. The vector $\mathbf{H}$ contains all unknowns of the system:

$$\mathbf{H} = \begin{pmatrix} h_1 \\ h_2 \\ . \\ . \\ h_n \end{pmatrix} = \begin{pmatrix} \mathbf{P} \\ \mathbf{W} \\ \mathbf{Q} \end{pmatrix} \text{ in case of the IK, or } \begin{pmatrix} \mathbf{P} \\ \mathbf{W} \\ \mathbf{X} \end{pmatrix} \text{ in case of the DK} \quad (7.6)$$

The next step is to create the Jacobian matrix of the system of equations. The Jacobian is defined as:

$$\mathbf{J}_{\text{nxn}} = \frac{\partial \mathbf{F}}{\partial \mathbf{H}} = \begin{pmatrix} \frac{\partial f_1}{\partial h_1} & \frac{\partial f_1}{\partial h_2} & \cdots & \frac{\partial f_1}{\partial h_n} \\ \frac{\partial f_2}{\partial h_1} & \frac{\partial f_2}{\partial h_2} & & \vdots \\ \vdots & & \ddots & \\ \frac{\partial f_n}{\partial h_1} & \cdots & & \frac{\partial f_n}{\partial h_n} \end{pmatrix} \quad (7.7)$$

7.2.2 Iterative Algorithm

This step consists in finding a solution of the system $\mathbf{F}$ for all unknowns $\mathbf{H}$, using a numerical iterative algorithm. The best known and most used scheme is the Newton-Raphson iterative algorithm:

$$\mathbf{H_{n+1}} = \mathbf{H_n} - \mathbf{J}_{\text{nxn}}^{-1} \cdot \mathbf{F(H_n)} \quad (7.8)$$

where $\mathbf{H_{n+1}}$ is the *new estimation for the solution*, $\mathbf{H_n}$ the *estimation from the previous iteration* and $-\mathbf{J}_{\text{nxn}}^{-1} \cdot \mathbf{F(H_n)}$ the *corrective term*. The Newton-Raphson algorithm is known as a very robust scheme that converges nearly always (48). Nevertheless, the very first estimation should be chosen close to a already known solution (e.g. using the CAD-model and measuring the first posture). The iterations are carried forward as long as the *corrective term* exceeds a certain user-defined threshold ε:

$$\mathbf{J}_{\text{nxn}}^{-1} \cdot \mathbf{F(H_n)} = \mathbf{H_n} - \mathbf{H_{n+1}} > \varepsilon \quad (7.9)$$

7.2 Geometric Modeling

The threshold ε is often chosen 100 times smaller than the expected precision of the machine. If after the k-th iteration the corrective term falls below the threshold ε the vector of unknowns $\mathbf{H_k}$ can be considered as solution to the system of equations $\mathbf{F}$.

Considering the complexity of computing an analytical Jacobian matrix, it is recommended to compute a numerical derivation for each element of the Jacobian:

$$\frac{\partial f_i(h_0)}{\partial h_i} \approx \frac{f_i(h_0 + \delta) - f_i(h_0)}{\delta} \quad (7.10)$$

where δ should be chosen very small to guarantee the validity of the upper equation. Considering computation time issues, the Jacobian being definitely the most consuming part, there exist different possibilities instead of calculating a new Jacobian at every iteration:

Fixed Jacobian $\mathbf{J_{nxn}(X_0)}$: The Jacobian is calculated once and in a certain point. Its values can be saved as constants and retrieved when needed. The center of the workspace $\mathbf{X_0}$ is often used to define a fixed Jacobian.

Local Jacobian $\mathbf{J_{nxn,j}}$: The workspace is split in different quadrants and a fixed Jacobian is attributed to each. The resulting set of Jacobian matrices is saved as constants. Depending on the relative placement of the robot's output in the workspace, a different Jacobian will be used.

Jacobian $\mathbf{J_{nxn}}$ calculated once per iteration loop: For each calculated point of the kinematic model a new Jacobian is generated, but not renewed for each iteration.

Depending on the computing power and the targeted precision, different methods might prove to be the best compromise.

The presented procedure allows to compute either the Direct Kinematics (DK) or the Inverse Kinematics (IK) depending on the declaration of the *known* and *unknown* variables. The above computation presented the articular coordinates $\mathbf{Q}$ as unknowns and the operational coordinates $\mathbf{X}$ as being known. It therefore represented the inverse kinematics. Declaring the variables the other way around and solving the system for $\mathbf{X}$ would create the direct kinematics.

7. ASPECTS OF INDUSTRIALIZATION

7.3 Optimizing Kinematics

Parallel/hybrid mechanisms generally possess a greater amount of geometric parameters than serial mechanisms. Their multiple kinematic chains naturally induce a higher amount of mechanical elements and geometric parameters. Now, for parallel/hybrid kinematics, the values of these geometric parameters are highly influencing the final performances of the robot (50)(48). This is why optimizing a kinematics is a determining and non negligible step in the design process of a robot.

When talking about *optimization* we need to stipulate which characteristics need to be optimized (*criteria*), and which goals and limits they need to lie within. As an example, a non-exhaustive list of often optimized criteria: stiffness, working volume, movement resolution, accuracy.

The project specifications of an industrial machine mostly confront us with an additional constraint: Some characteristics of the machine need to guarantee a minimum performance such as a minimal workspace/footprint ratio. In industrial reality, the optimization of a kinematics is therefore the combination of a multi-criteria optimization together with the retaining of minimum requirements on different characteristics.

In all cases a powerful modeling system is needed, whether it is a CAD system that represents the mechanism or the mathematical, geometric models presented in section 7.2 that allow computing the mechanism in all poses. Choosing the significant geometric parameters is also an important cornerstone of a correct model. The designer should keep the amount of parameters as low as possible by considering symmetry aspects, relative placement of elements and the achievable tolerances of the machine parts[1].

For simple cases – when the amount of geometric parameters is small – an *intuitive trial-and-error method* can be used. The limited amount of parameters makes it possible to figure out their influence on the criteria that has to be optimized. Figure 7.3 illustrates a simple kinematics with different sets of parameters. A CAD motion simulation allows to rapidly analyze the impact of the parameter modifications[2].

More complex cases need a different approach, as the interaction of the different geometric parameters becomes immensely complex. A very straight-forward and rapidly implemented approach was used to optimize the kinematics of the Alpha5 robot (see

---

[1] For example, the linear arrangement of axes can be considered as perfect for optimization tasks. For the control and calibration of the machine however, this might not be the case anymore.
[2] By measuring and plotting joint strokes, joint velocities, static forces on joints etc.

7.3 Optimizing Kinematics

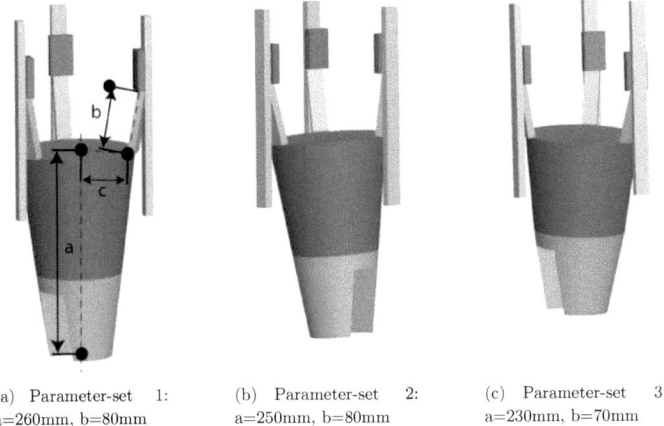

(a) Parameter-set 1: a=260mm, b=80mm

(b) Parameter-set 2: a=250mm, b=80mm

(c) Parameter-set 3: a=230mm, b=70mm

Figure 7.3: Example of 3 different optimal parameter-sets for differently sized working volumes (x/y/z), the biggest being on the left. The criteria to be optimized were the angular strokes of the pivots (in point P) connecting the 3 arms to the central platform, and a best possible resolution at the output ($\frac{c}{a}$).

figure 3.6). It consists in a *numerical method which tests all possible parameter-sets*[1] whether they fulfill the minimum requirements. The figure 7.4 illustrates a part of the source-code which generates the parameter-sets and evaluates their overall performance.

Once the parameter-set generated (upon entering the last loop) the programm will *process the whole working volume and evaluate the characteristics* of the corresponding mechanism. The processing of the working volume is delimited by a set of constraints, typically the limits of the joints. The results of the whole processing is a cloud of points which represents the working volume, and where each point has an attributed value that reflects the investigated characteristic. The modeling system in this case was the inverse kinematic model[2] of the robot.

Each set of parameters has then to be evaluated and classified using a *cost function* (sometimes also called fitness function). This function integrates all criteria that are

---
[1] Given certain design/footprint limits.
[2] See section 3.2.4 for definitions.

7. ASPECTS OF INDUSTRIALIZATION

```
for(l2_factor=0.5;l2_factor<3;l2_factor=l2_factor+0.1)
{
    l2=l1*l2_factor;
    for(l3_factor=0.3;l3_factor<1;l3_factor=l3_factor+0.1)
    {
        l3=l1*l3_factor;
        for(l4_factor=0.15;l4_factor<0.4;l4_factor=l4_factor+0.05)
        {
            l4=l1*l4_factor;
            for(Ra_factor=0.2;Ra_factor<1.5;Ra_factor=Ra_factor+0.1)
            {
                Ra=l1*Ra_factor;
                for(Rb_factor=0.2;Rb_factor<0.5;Rb_factor=Rb_factor+0.05)
                {
                    Rb=l1*Rb_factor;
                    for(fact=0.1;fact<0.6;fact=fact+0.05)
                    {
                        for(hauteur=0;hauteur<=9000;hauteur=hauteur+200)
                        {
                            rayon=0;
                            for(counter_interval=0;counter_interval<8;counter_interval++)
                            {
                                for(i=0;i<=140;i=i+20)
                                {
```

Generating Parameter Set

Processing through workspace

Figure 7.4: A numerical optimization program whose goal is to find the best possible parameter-set for a given criteria. The first few loops generate a parameter-set which describes the geometry of the "virtual" machine. Using the inverse kinematics, the whole workspace of the machine is processed and its performance evaluated in each point.

considered for the optimization. It can be more or less complex depending on the weighting or mathematical entanglement that one wishes to express. The following equation shows a simple cost function for the maximization of the working volume:

$$F = h\varnothing + h \quad (7.11)$$

where h is the measured height of the working volume and $\varnothing$ is the measured diameter. In this case a higher h is privileged over the diameter, therefore a bit stronger weighted. In general, a cost function can be expressed as the weighted sum of all examined criteria:

$$F = \sum_i w_i c_i \quad (7.12)$$

where c_i is the i-th inspected criteria and w_i its weighting coefficient.

After a first optimization run, the designer can choose the best parameter-sets (with the highest cost function) and restart an optimization, with more precise boundaries,

7.4 Protection

in order to refine the results. Figure 7.5 schematizes the method and illustrates the method, applied on the Alpha5.

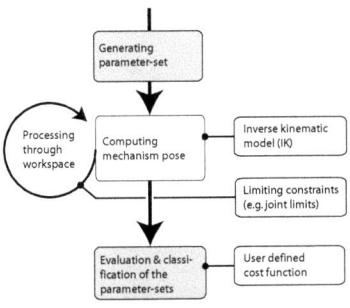
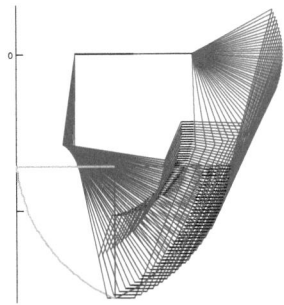

(a) Global Principle of the numerical optimization method.

(b) Section through the working volume of the Alpha5 which was generated using the numerical optimization method. The poses on the outer, right limit of the working volume were visualized in order to control the proper working of the programm.

Figure 7.5: Principle of the numerical optimization method and its application to the Alpha5.

By experience we can state that the intuitive trial-and-error method can be efficiently used till 4-5 geometric parameters (typically a 2-3 axes mechanism). An attentive designer can figure out the effects of the parameters and their correlation. For a higher amount, the use of a numerical approach based on a cost function is advised. This type of optimization method is the most widespread in the scientific community. Different methods using *genetic algorithms* (26)(10) have been developed in the last years, their results are promising, but the convergence of these algorithms cannot always be guaranteed.

7.4 Protection

Although machine tools procure the impression of being very robust, they have very delicate and sensitive elements that need protection. The nature of the harmful substances can be very different: impurities or chips from the processing of the workpiece, wear of tool or machine elements, cooling/lubrification liquids, particles which can infil-

7. ASPECTS OF INDUSTRIALIZATION

trate from the outside, sometimes even the ambient air! Although this is a very delicate subject for every machine tool industry, there exist barely no scientific literature which addresses this subject for parallel/hybrid mechanisms. Though, the problem is not the same as for serial kinematics.

Figure 7.6: Inside view of a machining center. Chips can be found everywhere, and will go everywhere you thought they wouldn't!

Polluting substances can be harmful for several reasons: One reason is that they can get jammed between 2 mobile parts (for example in joints) and exert very high constraints, by this influencing the precision of the machine or even creating damage. Another reason is that they can affect the quality of surfaces (corrosion, erosion). Since the machines are intended to work incessantly for years, the protection is taken extremely seriously. Among the sensitive elements we can find joints, motors/spindles, transmissions, cabling, sensors.

The protection is achieved by splitting the space around the clean, sensitive element and the polluting source by a mechanical interface. This is mostly a simple task for motors and transmissions that do not move, however for joints – which by definition provide movement between two parts – the problems becomes more difficult. The protection system for a joint needs to adapt to this movement.

There exist 2 different **protection technologies** to approach this problem (both technologies are illustrated each by an example in figure 7.7):

7.4 Protection

Articulated rigid elements, like telescopic metal covers, are articulated protections that can follow the movement of a joint. They are a composed of a series of simple flat elements. These elements form a mechanism, thus, each element needs to be correctly guided. The sealing between the elements is mostly guaranteed by scrapers and assisted by a slight overpressure in the clean part of the machine. Friction between the rigid elements is generated.

Flexible elements, like bellows or membranes, provide the necessary movement by elastic deformation (like flexible joints). In order to guarantee high strokes they are mostly made of multiple folding. This gives them the ability to achieve very complex movements if necessary.

(a) A collection of rubber bellows from Beakbane, a specialist for machinery protection (www.beakbane.co.uk)

(b) Example of a rigid telescopic cover that was developed by Beakbane. The cover protects a horizontal translational axis.

Figure 7.7: Two different protection elements representing the 2 approaches: flexible elements and rigid elements.

Rigid protection elements have a better durability since they are less susceptible to abrasive elements and aggressive liquids. It is common in the machine tool industry to periodically exchange flexible protection elements during the lifespan of a machine, whereas rigid elements can last the whole lifespan. Therefore, rigid protections are preferred.

Besides the protection technologies, we differ between 2 different **protection concepts** (both concepts are illustrated each by an example in figure 7.8):

Element-wise local protection: Protecting all sensitive elements apart.

7. ASPECTS OF INDUSTRIALIZATION

Global protection: Regrouping elements or protecting the whole mechanism all at once.

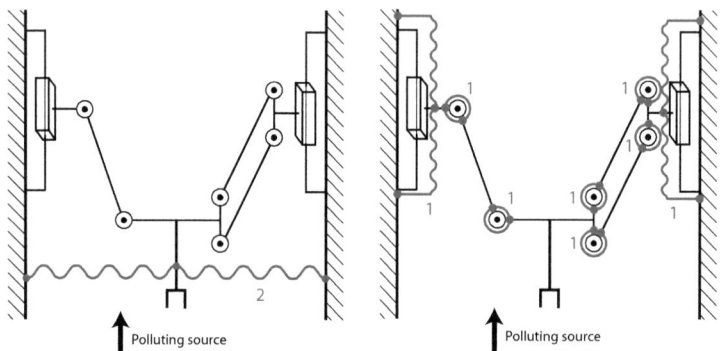

(a) Concept of global protection on a parallel 2 axes robot. The protection reaches from the base to the output of the robot.

(b) Concept of element-wise local protection on a parallel 2 axes robot.

Figure 7.8: The 2 different protection concepts. The numbers indicate the amount of DOF that the corresponding protection system needs.

The figure 7.9 presents different systems to protect a spherical/universal joint.

A protection element – whether it is global or local – needs the same type, and at least the same amount, of DOF as the mechanism which lies underneath. Considering this, we can draw different conclusions:

- A global protection system generally needs a high amount of mobilities.

- A global protection system *using rigid elements* can grow extremely complex for machines with high amount of mobilities, all the more if several tilting movement are present[1].

- As parallel mechanisms possess more internal DOF – more joints to protect – than their serial equivalents, a global protection can be very effective.

---
[1]Serial kinematics, in order to avoid this difficulty, always protect the tilting heads axis by axis. For translations however, they sometimes regroup 2 or 3 axes in a same protection system

7.4 Protection

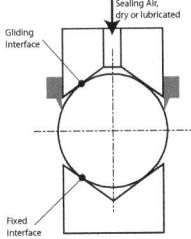 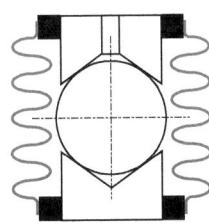 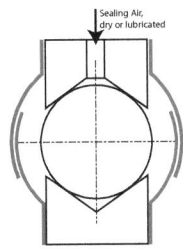

(a) The most compact solution consists in placing a hydraulic/pneumatic scraper directly on the joint interface. This type of scrapers is used for alternating back-and-forth movements in pistons, typically the movements obtained in a spherical joint.

(b) A protection system based on flexible bellows. The entire joint is covered and no interface is present. However, the system requires space.

(c) A protection system using articulated rigid elements. An exterior structure, which guarantees the same mobilities, is added and kept as closed as possible. The remaining space between the protection covers can be sealed with a scraper.

Figure 7.9: Protection concepts for spherical/universal joints, except solution (a) which is only viable on spherical joints.

The points above clearly show the difficulty of protecting parallel or hybrid mechanisms: It is economically more interesting to integrate a single, global protection system but its technically difficult to accomplish. As an (extreme) example, the Alpha5 which possesses a mobility inefficiency of 24.4 (chapter 5). A complete element-wise protection would require 24-times more mobilities than a global protection system.
Furthermore, the protection of high-mobility rotative joints, which proves to be sensitive, is typical for parallel and hybrid kinematics. In serial kinematics however, there are only translational joints (sliders) or 1 DOF rotative joints (pivots) to protect.
Recently, new synthetic, flexible materials have been developed and their durability in machine tools environment has to be verified on long-lasting wear tests. If their resistance proves to be enduring they may facilitate the integration of flexible, global protection systems.

The needed space to include the protection systems is often neglected during the first phases of machine design. Integrating a protection system consists in *adding a passive mechanism* that serves as an interface between the sensitive mechanisms and the polluting sources. Designers often neglect that this mechanisms needs the same

7. ASPECTS OF INDUSTRIALIZATION

strokes, which induces a minimum functional size. Therefore, a designer should consider the protection system as soon as possible in the design process. Some companies push this thought even further and underline that "they design the machine according to the protection system".

7.5 Cabling

The power- and signal-cables are connecting the electrical drives to the actuators. They need therefore to accomplish the same movement[1] as the motors they are connected to. This represents an issue that is taken very seriously in the machine tool industry. A badly executed cabling can rapidly lead to an error-prone machine, as the cables can get stuck in mechanisms, be folded below the minimum radius or simply worn-out due to constant friction.

Considering the cabling aspects of a machine tool, the parallel/hybrid mechanisms present a certain advantage over purely serial mechanisms. Section 4.2.4 already introduced the two principal methods of actuating a parallel kinematics. Both methods[2] present the advantage of having the actuators very early in the kinematic chain[3] which, besides the enhancing the dynamics of the machine, shortens the cabling distance and reduces the movements of the cables.

Figure 7.10 illustrates the common **cabling concepts for parallel kinematic chains**.
The resulting advantages on cabling issues are:

Short cabling distances: The cabling distances are reduced thanks to the motors being placed close to the machine frame.

Low movement: The movement to be performed by the cables is small, or even inexistant for mechanisms which move the base-point.

In **serial kinematics and in certain hybrid kinematics** – due to their stacked arrangement of the axes – the cabling has to pass through multiple axes before reaching the corresponding motor. This implicates a long cabling distance and a cumulation of movements that the cable has to perform during operation. A widespread concept is to

---

[1] With respect to the base/frame.
[2] Moving base-point and varying strut-length.
[3] Considering the machine frame as the beginning of a kinematic chain, they are very close to the machine frame.

7.6 Characterization Measurements

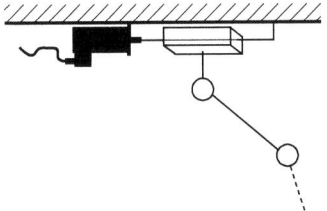
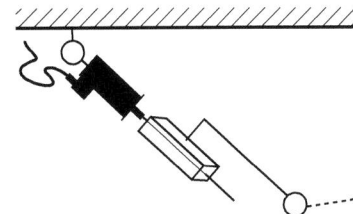

(a) Parallel kinematics with moving basepoints. Best case for cabling since the motor doesn't even move. The cables are fixed and can be easily protected.

(b) Parallel kinematic with varying strut-lengths.

Figure 7.10: Typical cabling of actuators on parallel kinematics. The two most common driving systems and their cabling are presented.

pull the cables through the parts and through the hollow shafts of the motors, thereby allowing the cables to work in torsion. This way, a more compact cable bending can be achieved. This solution has been applied to the Omikron5 presented in section 9.3 (figure 9.28).

7.6 Characterization Measurements

Once a mechanism is designed and put into operation it has to be characterized in order to validate the chosen mechanical concepts. There are 3 important and revealing measurements that can – with a optimized setup – be rapidly implemented: **Repeatability/Hysteresis, stiffness and the Eigenfrequencies.** These 3 characteristics, their consequences on the overall machine performance and their measurement setups will be explained in this chapter.

7.6.1 Repeatability

There is often confusion between the two terms *precision* and *repeatability*. Let's review the meanings first:

Precision (or accuracy) is the ability of a machine to precisely *reach a posture* given some input values. Good precision guarantees a short difference (error) between a

7. ASPECTS OF INDUSTRIALIZATION

targeted posture and the input. Calibration consists in establishing a relationship between measured postures and targeted postures, and aims to improve precision. Precision describes the ability of a machine to manufacture/handle in certain geometric tolerances.

Repeatability is the ability to *reproduce a posture* given some input values. Good repeatability guarantees a certitude in reaching the same posture over multiple cycles. However, it does not quantify the difference (error) between the effective posture and the targeted posture. The mechanical qualities of a machine greatly define the repeatability: backlash, stiffness, friction, hysteresis. The repeatability therefore informs very well about the quality of the mechanics without taking the calibration – a later step in the mechanism design – into account.

Figure 7.11 graphically explains the differences between both terms. For reasons of simplicity it has been applied to a 2-axes $(x-y)$ space.

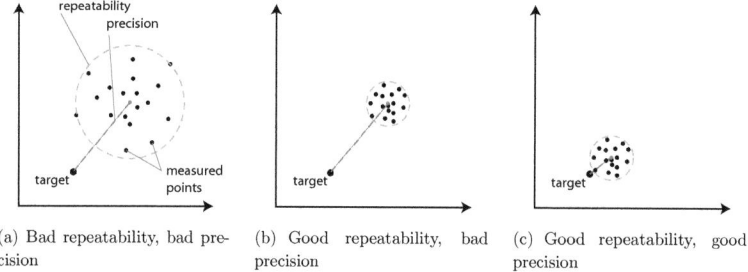

(a) Bad repeatability, bad precision   (b) Good repeatability, bad precision   (c) Good repeatability, good precision

Figure 7.11: Definition of repeatability and precision.

We therefore need to determine the repeatability of a mechanism in order to quantify its potential of being precise.

The proposed measuring setup consists of a cube which was machined to hold 3 inductive touchprobes. It is represented in figure 7.12. The 3 touchprobes are equipped with perfectly planar tips and are in contact with a sphere (the target). This configuration guarantees *a single contact point* and filters out any parasitic rotation. The target is attached to the end of the machine, replacing the tool. The cube is attached to the ground using a adjustable support[1].

---
[1] Which should be as stiff as possible and isolated from any exterior perturbation.

7.6 Characterization Measurements

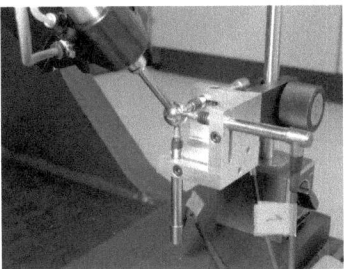

(a) Measuring setup applied on the Omikron5. The repeatability of the translations was analyzed. The cube is fitted on an adjustable support. This support needs to be stiff, so that no errors are introduced.

(b) Measuring setup applied on the MinAngle. The parasitic translations were analyzed in order to verify the mathematical model.

Figure 7.12: Applications of the measuring cube. The cube is composed of 3 orthogonally distributed inductive touchprobes. All of them have a perfectly defined contact with the target (in this case a sphere).

This setup was used to determine the translational repeatability of multiple mechanisms. A similar system can be used if a 6-axes measurement is needed. In this case – when rotations are measured – the central sphere has to be replaced by another shape, typically a cube, and the touchprobe-tips by a spheres[1]. This concept is illustrated in figure 7.13.

In order to statistically validate a measurement, there need to be several movement cycles (see figure 7.14). The mechanism needs to move the target into the measuring point and back again, and from different approach direction while moving all axes, conducting a position measurement at every cycle. This repeated measure should be done in several regions of the working volume, particularly in regions where strong non-linearities are expected (close to singularities).

Characteristics that influence the repeatability of a machine are backlash, stiffness, friction and hysteresis. Backlash is an uncontrolled movement between two mechanical elements and can randomly influence the position of the tool. Stiffness – if not high enough – can also influence the position depending on the robot's posture (with respect to gravity) or depending on externally applied forces (e.g. cutting forces). Hysteresis

---
[1] In order to guarantee exactly 6 precise contact points.

7. ASPECTS OF INDUSTRIALIZATION

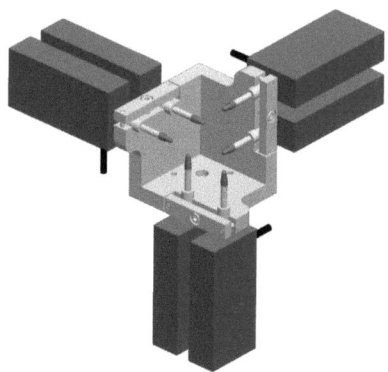

Figure 7.13: A 6-axes measuring cube. The target takes the general shape of a cube.

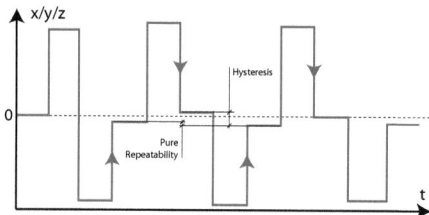

Figure 7.14: Measuring pure repeatability and hysteresis. Several measures are made in order to statistically validate the measurements. The sum of *pure repeatability* and *hysteresis* is denoted simply as *repeatability* in this work.

is a direction-depending phenomenon due to friction, and it can modify the position of the tool depending on the direction that was chosen to approach the position.

7.6.2 Stiffness

Stiffness is the resistance of a elastic structure to deflection or deformation by an applied force and is defined in section 3.1.1, equation 3.1.

The stiffness of the several prototypes was measured using the same instrument presented in section 7.6.1. The measuring cube is centered in the examined point and

7.6 Characterization Measurements

an external force is applied to the mechanism. Different applied forces and the corresponding displacement measurements will generate a graph. The numerical derivation of this graph provides the stiffness.

In contrast to serial mechanisms, the parallel/hybrid mechanisms can exhibit extremely varying results depending on the examined position in the working volume. Positions close to singularities usually represent worst-case scenarios.

While measuring the stiffness it is very important to be conscious of the measuring- and force-loop. The stiffness of all elements **where both loops are coincident** will be taken into account. A typical error is to direct the force through the measuring setup, by this, taking the stiffness of the measuring setup into account. Figure 7.15 illustrates a correct and a wrong example of directing the force loop.

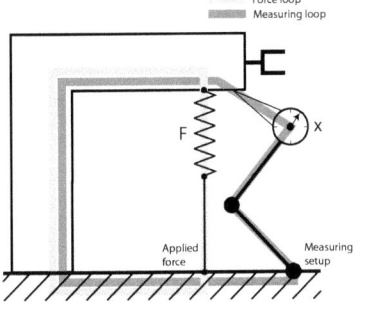 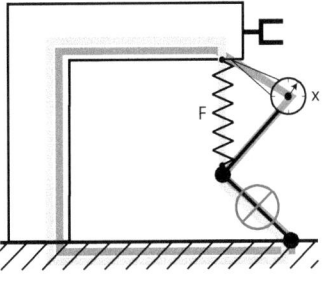

(a) Correct directing of force- and measuring-loop.

(b) Wrong directing of the force loop. The measured stiffness will include the bending of the adjustable support.

Figure 7.15: Stiffness measuring setup and the influence of the force- and measuring-loop.

7.6.3 Eigenfrequency

The 1st Eigenfrequency is the 1st vibration mode of a mechanical structure. As explained in section 3.1.4 it is crucial for high-precision and high-speed machines to place it as high as possible.

The Eigenfrequency can be measured by placing an accelerometer on the output of the robot and exciting the vibration modes. This excitation, a nearly perfect Dirac

7. ASPECTS OF INDUSTRIALIZATION

impulse, is made by hitting the output (e.g. a hammer that is simultaneously triggering the measure). Modern oscilloscopes can then compute the FFT (Fast Fourier Transformation) of the recorded accelerations[1] and determine the frequency response. The 1st Eigenfrequency of the robot is represented by the 1st peak in the amplitudes. Figure 7.16 represents the frequency response of the MinAngle robot.

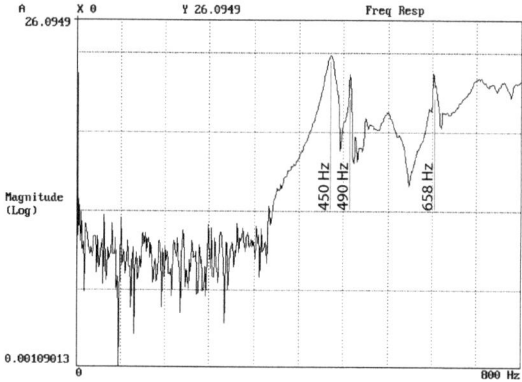

Figure 7.16: The frequency response of the MinAngle's left hand to a mechanical excitation (43). The first eigenfrequency is located at 450Hz, visible because of the peak.

7.7 Summary and Conclusion of the Chapter

This chapter presents aspects of hybrid kinematics which, besides the pure mechanical aspects presented in chapter 6, contribute to the performance and industrialization of such mechanisms. All the points presented in this chapter complete the designer's vision on the differences between serial and parallel/hybrid kinematics.
The distribution of DOF is presented and its positive impact on the mechanical performances is pointed out. Different distribution concepts are cited. A general, numerical approach for the mathematical modeling is provided and its profit for the optimization task is shown. A straight-forward optimization method based on cost functions is presented.
Protection and cabling issues/concepts are presented. The difficulty of their integration

---
[1]Including their directions in space.

7.7 Summary and Conclusion of the Chapter

is shown, and the difference to conventional machine tools is pointed out. The protection of hybrid mechanisms proves to be more difficult than for conventional machines, the cabling however easier.

Final characterization measurements are listed, and a general measuring system therefore is presented. These measurements allow the verification of the machine's performance and, thereby, the validation of chosen concepts.

All cited aspects of machine design are integrated in a design methodology. This methodology is outlined in the following chapter and used on concrete examples in chapter 9.

7. ASPECTS OF INDUSTRIALIZATION

8
Design Methodology

This chapter presents a stepwise, iterative methodology to design hybrid-kinematic mechanisms. The methodology proposes a series of steps which should assure the best possible adaptation and optimization of the mechanisms considering the application specifications. It emphasizes certain steps which are often neglected by designers. Generally, the earlier the steps in the design progress, the more important and influential they are.

Machine design is not an exact science. It is based on catalogues of solutions, objective analysis, experiences made and design iterations. This methodology takes this into account.

8.1 Global View

The methodology is illustrated in figure 8.1 and is divided into 3 parts:

Left part:	Central part:	Right part:
The left part indicates inevitable, significant loops in the design process. These loops stand for an iterative process of adaptation and optimization. They indicate an eventual point for continuation if the iterative process failed.	The central part provides, in chronological way, the different steps in the machine design process.	The right part proposes elements presented in this thesis and which can be helpful at this very moment of the design process.

Table 8.1: Structure of the design methodology.

8. DESIGN METHODOLOGY

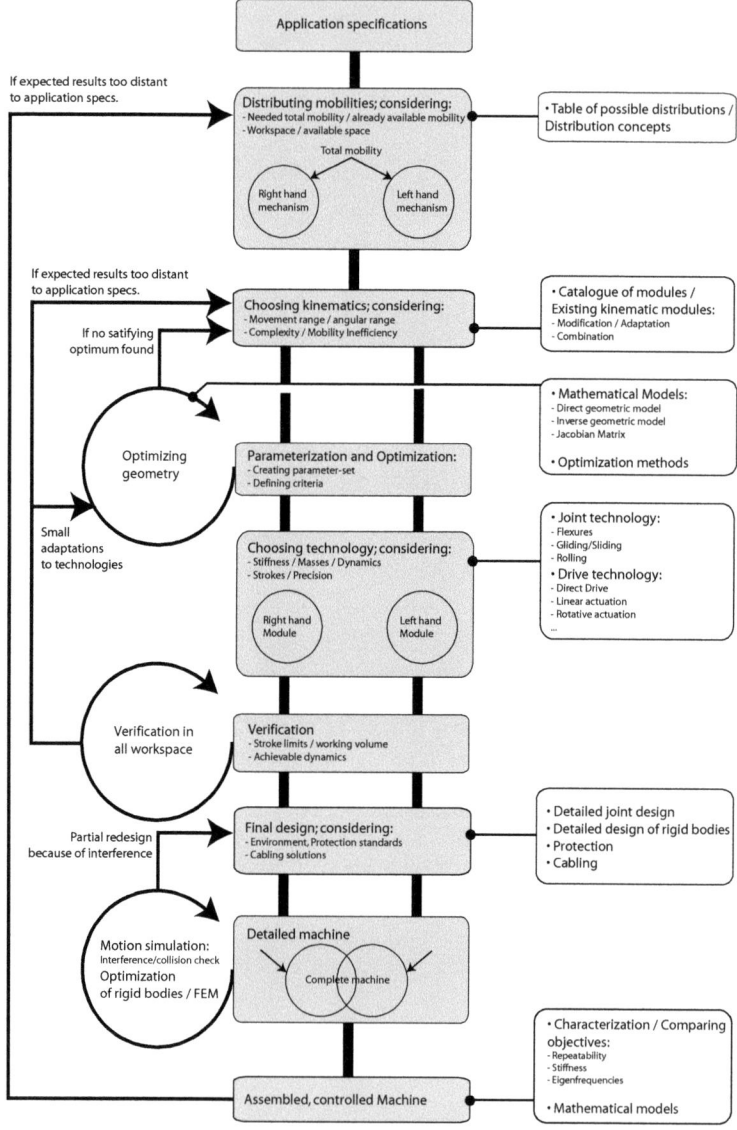

Figure 8.1: Design methodology for hybrid mechanisms.

The CAD systems nowadays available offer the very beneficial ability of modeling **mechanism motion**. This ability allows different simulations and verifications that would otherwise be very time-consuming, especially for parallel/hybrid kinematics, because of their non-intuitive movements and their high risks of collisions. It is therefore absolutely recommended to keep an *updated CAD model of the machine at every step of the machine design*. The steps in the machine design process are detailed in the next section.

8.2 Stepwise Description

This section provides a stepwise and chronologically classified description of the single designs steps, according to the methodology presented in figure 8.1.

8.2.1 Application specifications

This step is about the analysis of the process and its interpretation[1] into machine performance. A lot of importance should be attached to this step as it sets the benchmark of all characteristics and future decisions.

Several different aspects have to be determined (63):

Technical performances: motion types, working volume (linear strokes, rotative strokes), process quality (stiffness, eigenfrequency, precision).

Economic performance: Throughput, production costs, energy consumption, costs of consumables.

Integration in shop floor: Compatibility in production chain (same control hardware/software, sharing subsystems like e.g. chip removal and lubrication), management of blank piece magazine.

8.2.2 Distributing mobilities

The distribution of the mobilities on the left and right hand is carried out. The tables in section 7.1 present all different distribution possibilities for 6, 5, 4 and 3-axes mechanisms. There are multiple thoughts that can restrict these possibilities (see section 7.1 for more details):

---

[1] As the word implies, there are multiple possible outcomes of this step. All should be kept in mind and retrieved whenever a decision has to be taken in the design process. For example the drilling of a wide hole: It can be achieved with a drilling tool of the same diameter, or by a circular toolpath interpolation of a thin tool. One interpretation might require a tool change whereas the other requires the movement of 3 axes.

8. DESIGN METHODOLOGY

Constraints limiting the distribution

- The workpiece or the tool cannot be moved, there is no choice in the distribution of mobilities. Certain applications require a fixed tool or workpiece, whether it is very sensitive to movement or because multiple right hands are interacting at the same time on the workpiece (the workpiece's movement shouldn't be synchronized with 1 module alone, forcing all other modules to stop machining).

- Available space. Limited available place around one of the hands limits the amount of DOF that can be integrated.

- Already available mobilities. The mechanism provides an enhancement to an already available machine. Some mobilities are already available, new ones will be added.

Distribution concepts

- The translations and rotations are on separate hands. Due to the different nature of translational movement or rotational movement, it is mechanically meaningful to separate these different movements on the 2 hands and create dedicated designs.

- Modularity or symmetry. It is reasonable to perfectly divide the DOF on the 2 hands, by this creating 2 equivalent or symmetrical modules[1].

- Separating axes with different characteristics/Separating different technologies. If there exist a big imbalance in wanted axis properties (e.g. high/low dynamics, strokes, velocities, stiffness, precision/resolution) or axis technologies[2] it is reasonable to separate them.

8.2.3 Choosing kinematics

This step of machine creation is one of the most influencing, and decisive, as it greatly influences the performance of the machine. As already mentioned in the introduction, this step mostly relies on existing kinematics. The designer can form a specific catalogue by choosing (from cited references, cited kinematics or the catalogue in appendix D) potentially matching kinematics. Relying on this catalogue the designer will modify

---

[1] For applications which require an odd number of DOF it might even be meaningful to increase the total DOF to an even amount. Besides generating a redundant axis it gives the possibility to create 2 identical modules.

[2] It is very probable, but not necessary, that highly differing axis properties induce a different technology.

8.2 Stepwise Description

or even create modules that match the application specifications. The vast choice of kinematics will already be slightly limited as the distribution of mobilities was made before. For each hand of the machine, a kinematics will be chosen. The kinematics are analyzed using different criteria:

- Amount of DOF and motion type

- Movement range: Linear stroke / Angular stroke

- Complexity / Mobility Inefficiency

- Mechanical potential: Stiffness / Dynamics etc.

- Required space for integration (size and shape)

As this step proved to be very determining in the design of the machine tool, it is recommended to build up a small catalogue of kinematics which meets the previously mentioned criteria, and then, to perform a detailed study of these kinematics by weighting their capacity with respect to all criteria.

8.2.4 Parameterization and optimization

Once the kinematics chosen, the optimization (see chapter 7.3) and customization of the mechanism can start. The kinematics has to be parameterized, all decisive geometric sizes have to be interpreted as parameters, and an optimization criteria has to be formulated.

These criteria are mostly expressed as ratios, for example workspace/footprint ratio, but can be more complex expressions depending on the amount of examined characteristics of the machine. If no acceptable parameter-set is found after several optimization attempts the designer is redirected to a former design step. If this happens, it is very probable that the kinematics does not fit the project requirements or does not cope with the constraints.

8.2.5 Choosing technology

After choosing and optimizing the kinematics of the module/modules, the designer starts choosing the technologies. Depending on the targeted dynamics, resolution and stiffness he will have to choose actuators, reduction ratios, sensors and joints. Chapter 6 presents some elements which can be taken into consideration.

8. DESIGN METHODOLOGY

8.2.6 Verification

This step is about the simulation and verification of the module's theoretical performance. Dynamic models of the modules are derived and the maximum forces/accelerations are compared to the application specification. The reduction ratios are adapted to the dynamics and the maximum achievable velocities are verified. Considering the targeted stiffness the joints and drive systems (e.g. ballscrew diameters, ballscrew nut) are chosen.

Some technological choices may lead to a reconsideration of geometric parameters: Flexible joints for example have a very limited stroke. However, except this drawback, their implementation is recommended (see section 4.2.2). A revised optimization of the geometric parameters might be necessary to reduce the maximum strokes on certain joints for example.
If the results do not meet the application specification, other kinematics have to be considered.

8.2.7 Final design considerations

Final design considerations that take into account: protection, cabling, sealing air support, lubrication systems, stroke delimiters, references for axis initialization etc. All secondary elements that are necessary for the correct operation of the machine are planned, designed and integrated.

8.2.8 Detailed machine design

Once all the above-named steps successfully resolved, the detailed machine design can take place. The final verification consists in *a complete interference/collision check over the whole workspace* (illustrated by an example in figure 8.2). The different parts of the mechanisms get their geometrical shape depending on their load cases (the loads they will be exposed to, section 6.3). A static analysis of the mechanisms is necessary, first using simplified analytic considerations (section 6.3) and then refined using FEM[1] simulations.

---

[1] Finite Element Method.

8.2 Stepwise Description

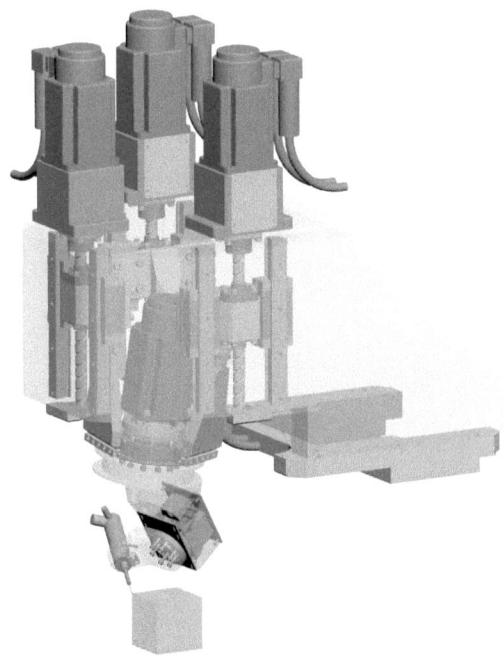

Figure 8.2: Omikron5 during final detailed design. The cube at the robot's output symbolizes the working volume and was used as a reference for a complete interference/collision check. The protection elements are not displayed in order to investigate the movement of all machine elements.

8.2.9 Assembled, controlled machine

The machine is assembled and the DK and IK have been implemented in the controller. Security limits (working volume, max. position/velocity errors)[1] have been integrated. The first movements of the machine are done with dismounted kinematics, the coupling between the axes is removed. This allows a verification of the correct working of DK/IK and security limits without the risk of a collision.

Characterization measurements are made and compared to the application specifica-

---

[1] On operational coordinates and articular coordinates. In contrast to cartesian machine tools they do not correspond.

8. DESIGN METHODOLOGY

tions. An analysis of the differences is carried out.

8.3 Summary and Conclusion of the Chapter

Machine design is a complex process which has to consider a lot of aspects and their interactions. Therefore, an iterative methodology is proposed. The multiple iterations help the designer understand the links between the different design aspects and allow a stepwise approach to a design optimum.

A big importance should be attached to the creation of vast catalogues of solutions for every design step. Especially the early steps of the process have to be catalogued rigorously as they have a high impact on the following steps, and therefore, on the overall performance of the machine. At an early stage several different concepts can be chosen (e.g. different distributions of mobilities, different kinematics), and a simultaneous execution of the design methodology, for each concept, can be performed.

For each design step this work proposes solutions or concepts. They are principally given in chapters 5, 6 and 7. The proposed design methodology has been applied to 3 case studies, all of them being subject to industrial constraints. Their development is summarized in chapter 9.

9
Case Studies and Prototypes

The following case studies present 3 hybrid-kinematic machines developed during the thesis and were chosen because of their contribution to this work. Their development was conducted using the methodology presented in chapter 8 and their corresponding sections will be structured the same way. Furthermore, the designer will draw profit from the experience gained during these developments.

9.1 Stewart3: Part of a hybrid 5-axes Machine for wire-EDM

9.1.1 Introduction

The Stewart3 mechanism was designed in collaboration with *Mecartex SA*[1] and *GF AgieCharmilles*[2] as a supplement to an existing 3-axes serial-kinematic translator, in order to expand its possibilities by adding 2 rotative axes and another redundant translation. The combination of both mechanisms forms therefore a complete 5-axes machine-tool (with an additional redundant Z translation) for wire-EDM. The additional 2 rotations will allow the machining of more complex features like mould draft angles, conical holes and oriented holes (fuel injectors).

The Stewart3 mechanism has to be implemented in an existing machine, therefore only very few and restricted space is available. The mechanisms will be completely submerged in the dielectric fluid and therefore requires a hermetic encapsulation.

The mechanism is represented in figure 9.1.

---

[1] www.mecartex.ch
[2] www.gfac.com

9. CASE STUDIES AND PROTOTYPES

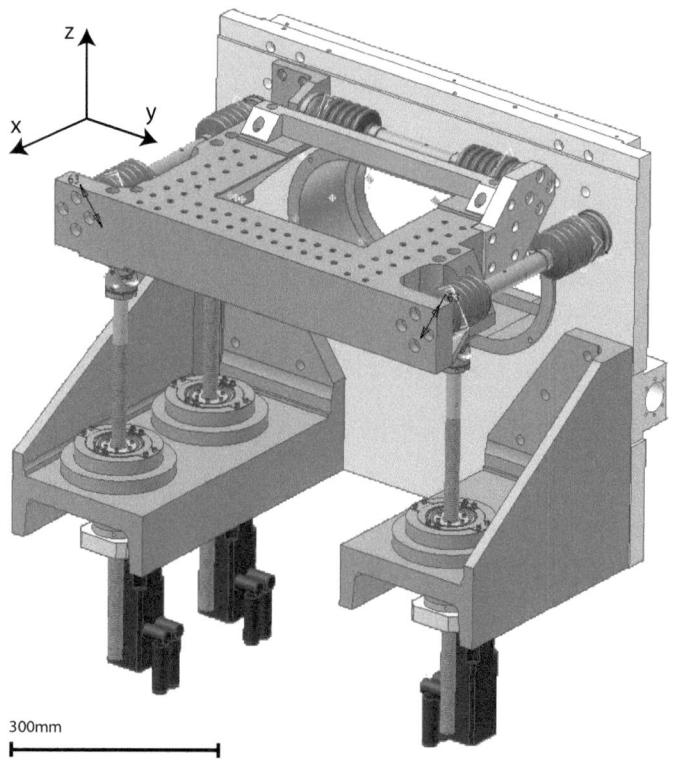

Figure 9.1: The Stewart3, a 3-axes orienting device for high-precision and high-stiffness applications.

9.1 Stewart3: Part of a hybrid 5-axes Machine for wire-EDM

9.1.2 Project Requirements

The project requirements are summarized in table 9.1.

Table 9.1: Summary of the Stewart3 project requirements.

Characteristic:	Value:
Total mobility	5 (3T2R), 3T already present
Linear stroke (z)	$\pm 50 mm$
Angular stroke (θ_x/θ_y)	$\pm 15°/\pm 15°$
Stiffness@Output (x, y, z)	$100 \frac{N}{\mu m}$
Angular repeatability (θ_x, θ_y)	72 arcsec
Integration aspects	Mountable on an existing machine, with a minimum of modifications, and fitting in a very shallow volume.
Protection aspects	Mechanism will be located in the dielectric fluid. Hermetic protection required.

9.1.3 Distribution of Mobilities

The mechanism which guides the wire and generates its feed[1] is completely implemented on the already existing translator. As this system is highly sensitive, it was decided to implement all additional DOF on the left hand, this way not modifying the right hand at all. The left hand, the table holding the workpiece, was fixed till now.

9.1.4 Kinematics

A very large catalogue of solutions was established and the different solutions were weighted regarding different aspects like stiffness, manufacturing costs, potential precision and the ease of implementation. At the end, a variant of the well-known Stewart Platform with only 3 active struts was chosen. Its 6 struts can be placed in a nearly optimal way, even in this very restricted volume (see figure 9.2). The orthogonal placement avoids that the struts enter the forbidden zone, in the middle of the volume. Furthermore, when using 6 struts of this type, only traction/compression efforts occur in the mechanism. A high stiffness can be guaranteed with such load cases.

The redundant Z-translation proved to be necessary. In fact, the quite large size of

---

[1] In order to achieve a high quality and precision of the wire-cutting process, the wire has to be continuously moved. During the machining the wire is slightly eroded and needs to be replaced.

9. CASE STUDIES AND PROTOTYPES

the table and its rotation around the horizontal axes θ_x and θ_y generate a considerable vertical movement at its borders. This movement has to be compensated[1]. Furthermore, the addition of this high-stroke translation allows to lift the workpieces out of the dielectric fluid. This avoids emptying the tank and grants an easy access to the workpiece.

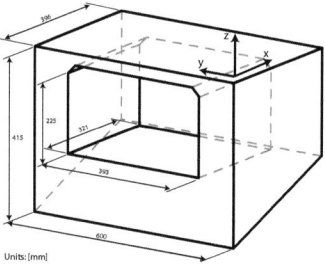
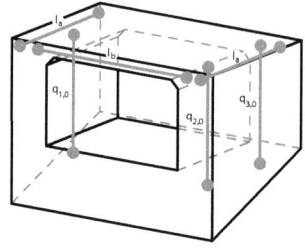

(a) Available volume. The table, on which the workpieces is clamped, is located on the upper horizontal surface of the volume.

(b) Placement of the 6 struts in the two lateral walls and in the upper surface. A perfect orthogonality can be achieved.

Figure 9.2: Available volume for the integration of the Stewart3.

The kinematics of the Stewart3, and of its translator, are represented in figure 9.3.

The whole machine presents a mobility inefficiency of 7.2 (total mobilities of 36, and 5 effective mobilities of the end-effector). This value is standard for all variants of Stewart platforms.

9.1.5 Optimization

The optimization of this mechanism can be reduced to a single problem: *Minimizing the parasitic displacements of the table $(x/y/\theta_z)$*. This directly reduces the movement amplitudes which has to be compensated by the translator. This way, less stroke is "wasted" just for compensation purposes.

In order to minimize the parasitic displacements the *lengths of the passive guiding struts should be maximized*. In the most extreme case, when the lengths of the guiding struts l_a, l_b (see figure 9.2, right side) tend towards infinite, the attachment points on the table

---

[1] The already present Z-translation of the right hand had not enough stroke for this.

9.1 Stewart3: Part of a hybrid 5-axes Machine for wire-EDM

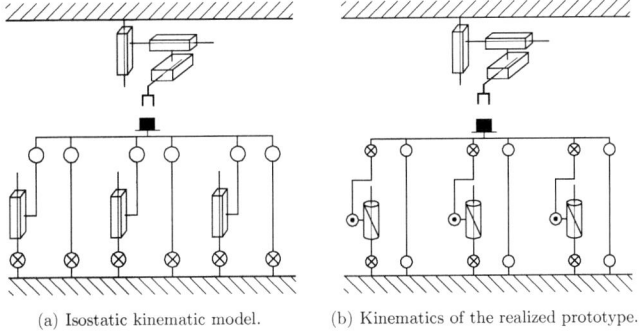

(a) Isostatic kinematic model.　　(b) Kinematics of the realized prototype.

Figure 9.3: The kinematics of the Stewart3 and of its translator.

can only move in a perfectly vertical direction. For this reason all spherical joints of the passive guiding struts were placed at the limits of the volume, this way leading to the maximum strut-lengths.

The stiffness of the struts (structural part) is linearly decreasing for longer strut-lengths, but as the expected values are very good, this decrease was accepted.

9.1.6 Technology

This section will be separated in two subsections: The 3 passive guiding struts, which are unactuated kinematic chains, and whose goal is to apply 1 constraint each on the table, and the 3 active struts. They differ in the choice of joint and preloading system.

Passive Guiding Struts

Several struts were proposed and evaluated. They are represented in the figures 9.4, 9.5, 9.6 and 9.7.

Finally, the first version using 2 separated spherical joints was chosen. The main reason being the smaller required space around the strut. In fact, compared to the other strut variants, no additional tube is lead between the 2 joints. Unlike represented in figure 9.4 it was decided to use the third-evolution spherical joints which have the particularity of having 2 working modes in a same design. This way, for the passive guiding struts which require spherical joints, as well as the active struts (see next section) which require universal joints, a unique joint design can be used.

9. CASE STUDIES AND PROTOTYPES

Figure 9.4: Simple bar with 2 separated first-evolution spherical joints.

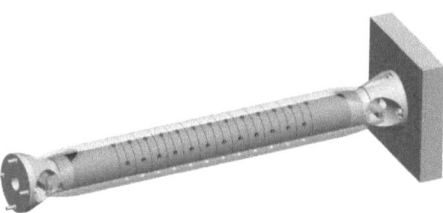

Figure 9.5: Strut with 2 modified first-evolution spherical joints sharing a common preloading system. This system consists of an elastic tube which has been partially cut, in order to obtain the required preloading force.

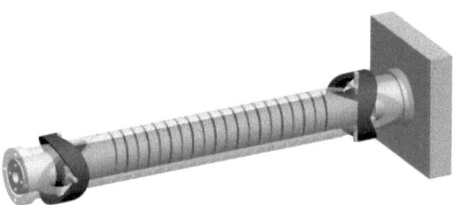

Figure 9.6: Strut with 2 modified third-evolution spherical joints sharing a common preloading system. The spherical joints contain unmodified bearing balls. The preloading system, an elastic tube, is mounted between the 2 universal joints.

The passive guiding struts possess an internal DOF (rotation of the strut around its main axis) which results from using 2 spherical joints, one at each end. If this proves to be problematic, one of the joints can be switched to universal joint mode.

9.1 Stewart3: Part of a hybrid 5-axes Machine for wire-EDM

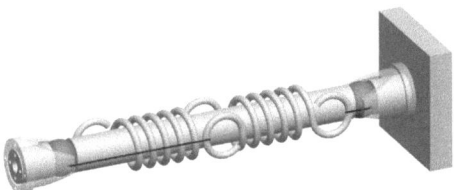

Figure 9.7: Strut with 2 modified third-evolution spherical joints sharing 2 identical preloading systems. The preloading system is composed of 2 standard springs which are mounted (using cables or rods) on each side of the joints.

Active Struts

The actuation is achieved by varying the strut-length, carried out by a ballscrew system. The ballscrew is mounted between two third-evolution spherical joints, but *working in universal-joint mode* (see section 6.2.3). The joints therefore have 2 DOF, the torsion being disabled. The whole set-up, its working principle, is represented in figure 9.8. The nut, which is mounted on a pivot, is actuated by a conventional rotative servomotor through a belt pulley system.

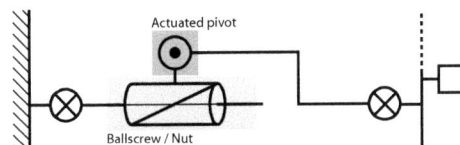

Figure 9.8: Kinematic scheme of a ballscrew drive providing a varying strut-length. The pivot, in reality aligned with the ballscrew, actuates the nut. The universal joint on the left is hollowed-out in its center, and placed exactly around the nut. For reasons of clearness they were placed behind each other in the figure.

In figure 9.8, the leftmost universal joint dimensioned very large and hollowed out in order to allow the ballscrew to pass through its center. Thanks to this the motor can be placed very close to the joint. This guarantees a compact design and avoids, when the strut tilts, a large lateral displacement of the motor.

The protection is achieved by placing bellows around each spherical and universal joint (some of these protections are illustrated in figure 9.1). The motors are located

9. CASE STUDIES AND PROTOTYPES

behind a separating barrier and are therefore completely isolated from the dielectric fluid.

9.1.7 Results and Conclusion

Table 9.2: Summary of the Stewart3's **expected** characteristics. The stiffness and the eigenfrequency only takes into account the struts and their single elements. The table as well as the machine frame was not taken into account as their design was not yet accomplished.

Characteristic:	Value:
1st Eigenfrequency	$300 Hz$
Angular stroke (θ_x/θ_y)	$\pm 15°$
Linear stroke (z)	$\pm 50 mm$
Max. parasitic displacement	$\pm 15mm$ (x), $\pm 17mm$ (y)
Stiffness linear $(x/y/z)$	$300/130/252 \frac{N}{\mu m}$
Stiffness angular $(\theta_x/\theta_y/\theta_z)$	$12 \cdot 10^6 / 1.7 \cdot 10^6 / 25.4 \cdot 10^6 \frac{Nm}{rad}$

The machine, while authoring this report, is in finally assembly stage and therefore unfortunately, no measurements could have been done yet. However, the performance of most elements, particularly the joints, has been proven in a different prototype (see Omikron5, section 9.3). The expected, computable performance of the Stewart3 machine are summarized in table 9.2. The final performance will strongly depend on the design of the table, as it has large dimensions and bears torsion and bending efforts. From a kinematics point of view the chosen parallel-kinematic mechanism proved to fit well in the very restricted, shallow space. A serial-kinematic mechanism, by definition composed of 1 kinematic chain, could not exploit the available space as efficiently as the Stewart3. Its stiffness would be lower and insufficient.

9.2 MinAngle: A hybrid 5-axes High Precision Machine Tool for μEDM

9.2.1 Introduction

The MinAngle concept (61) has been developed in collaboration with *Mecartex SA* and *GF AgieCharmilles* to satisfy needs in the domain of μEDM. The motivation is the same as for the Stewart3 machine (section 9.1), but on another scale. The two projects differ in the size of the machine and workpieces. The goal is therefore to design a 5-axes machine of small size, for small parts, with an even higher repeatability and precision. The developed prototype is represented in figure 9.9.

9.2.2 Project Requirements

The project requirements of the MinAngle are summarized in table 9.3.

Table 9.3: Summary of the MinAngle project requirements.

Characteristic:	Value:
Needed mobility	5 (3T2R)
Linear stroke (x/y/z)	$\pm$5mm
Linear repeatability	$\pm$5nm
Angular stroke (θ_x/θ_y)	$\pm$15°
Angular repeatability (θ_x, θ_y)	0.1 arcsec
Machine footprint	max. $\varnothing$500mm

The required repeatability clearly diverge from classic machine design. In such an order of magnitude of repeatability, the use of flexure-based joints is firmly advised (35)(36). For this reason the joint technology was chosen from the beginning. The main challenge of this project became the development of a fully-flexure based kinematics, and additionally, finding a way to exceed the angular limits of flexure.

9.2.3 Distribution of Mobilities

The decision of decoupling the translations from the rotations was rapidly taken. This decision simplifies the control and enables to design dedicated kinematics, which was

9. CASE STUDIES AND PROTOTYPES

Figure 9.9: The MinAngle μMachine. A complete 5-axes machine tool for high-precision applications.

necessary in order to obtain large angular strokes. Furthermore, a very effective kine-

9.2 MinAngle: A hybrid 5-axes High Precision Machine Tool for μEDM

matics for translation was available: The DeltaCube (40)(8)(36). This mechanism was also developed at the Laboratoire de Systèmes Robotiques (LSRO) and commercialized in its 4th version by Mecartex SA. To guarantee very high performances, in terms of precision and repeatability, this module is also completely composed of flexible joints. The DeltaCube was designed to carry a μEDM tool, therefore the planned mechanism on the left hand would carry the workpiece. A kinematics with 2 rotative DOF (θ_x/θ_y) had to be invented or developed.

As the DeltaCube is already well known, characterized, and completely fulfilling the requirements of the translational right hand, it won't be analyzed anymore in this chapter. The following sections will therefore focus on the novelty of the MinAngle concept, namely its left hand[1].

9.2.4 Kinematics

As decided in the precedent design step the left hand mechanism would need 2 rotational DOF. Starting from this we examined existing parallel kinematic mechanisms, because of their potentially higher precision and stiffness compared to their serial equivalents. Two similar mechanisms attracted our interest, the Dunlop and Orion kinematics (represented in figure 9.10).

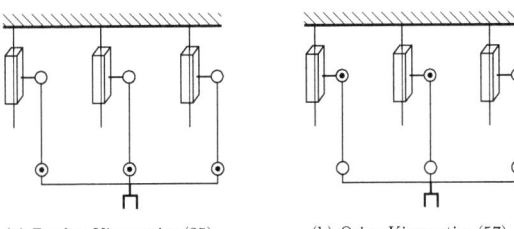

(a) Dunlop Kinematics (25).   (b) Orion Kinematics (57).

Figure 9.10: Dunlop (25) and Orion (57) kinematics. Two 3-DOF (1T2R) parallel mechanisms.

Both mechanisms are similar and possess 3 DOF, which is more than requested by the project requirements. Regardless of this, it has been decided to continue the development based on these two kinematics. The main reason being that the additional,

---
[1] The term MinAngle stands for the whole 5-axes concept. The newly developed kinematics of the left hand will be designated as *MinAngle's left hand*.

9. CASE STUDIES AND PROTOTYPES

and redundant, Z-translation may be useful for a better access to the working volume.

The critical point was to adapt these kinematics[1] in order to achieve the required rotation amplitudes of ±15° at the output. For flexure-based mechanisms this represents a very high amplitude which was never accomplished yet. The basic idea was to modify the kinematic chains in order to equally divide their two output angles[2] (θ_x and θ_y) on the different pivots constituting them.

Dividing θ_x on several pivots

First of all, the needed spherical joints (see figure 9.10), which are very difficult to realize using flexures, were replaced by 3 pivots that generate a very similar movement. One of these pivots, the one responsible for θ_x rotation, was then replaced by a four-bar linkage that has the ability to split an angle equally on the 4 pivots it is constituted of. Figure 9.11 resumes the modifications made.

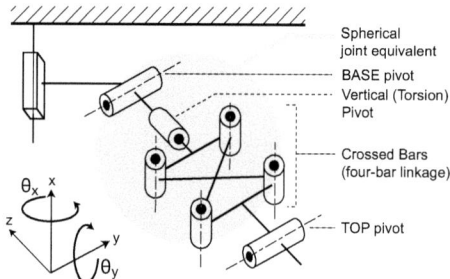

Figure 9.11: Kinematic chain of the MinAngle's left hand. Due to their spacial arrangement, each pivot can be attributed to a movement of the output. The pivots in the crossed bars can generate rotation around θ_x, and the "top" and "base" pivot generate rotation around θ_y.

---

[1] The adaptations that were made can be applied to both kinematics (Dunlop and Orion), therefore, there is no need to restrict our choice to one.

[2] The angle performed by the last segment of the kinematic chain. As they are rigidly coupled, this segment and the end-effector perform the same movements.

9.2 MinAngle: A hybrid 5-axes High Precision Machine Tool for μEDM

Dividing θ_y on several pivots

The modified kinematic chain now possesses 2 pivots to execute the θ_y rotation ("base" and "top" pivots in figure 9.11). The $\pm 15°$ on θ_y can be split on these two pivots by means of an optimization of the geometric parameters.

The resulting kinematics divide the $\pm 15°$ on all pivots (61). The single joints never perform more than $\pm 7.5°$. Compared to other systems this is not obtained through special materials, or complicated joint design (39), but only thanks to the kinematics. Thus, we do not decrease the stiffness nor the precision of such a system.

Figure 9.12 shows the complete kinematic scheme of the MinAngle concept. The left scheme illustrates the isostatic version, whereas the right scheme illustrates the finally realized overconstrained kinematics. Overconstraining a mechanism is normally not recommended, but in this case, because of the monolithic fabrication using EDM and the resulting geometric precision of the kinematic chains, this does not affect the operation of the mechanism.

RCM Capability

Since both mechanisms – right and left hand – are flexure-based they feature quite limited strokes. Fort his reason, it was an important aspect to create a kinematics that does not induce high parasitic translations when rotating. This way, the restricted linear stroke of the DeltaCube can be used profitably. This characteristic, called RCM (Remote Center of Motion, see figure 9.13 for an example) was successfully integrated in the MinAngle kinematics.

Unfortunately, the MinAngle kinematics does not present a perfect RCM, hence a little parasitic displacement is always present. However, the amplitude of this unwanted movement is reduced by optimizing geometric parameters.

The resulting mobility inefficiency of the complete MinAngle amounts to 16.8 (total mobilities of 84, and 5 effective mobilities of the end-effector) for the isostatic version, and 12 (total mobilities of 60, and 5 effective mobilities of the end-effector) for the realized, overconstrained variant. Both values are high. The reasons are multiple: The complexity of the kinematics is increased with four-bar linkages in order to allow high angles (crossed bars). Additionally, since there exist no basic flexures to create

9. CASE STUDIES AND PROTOTYPES

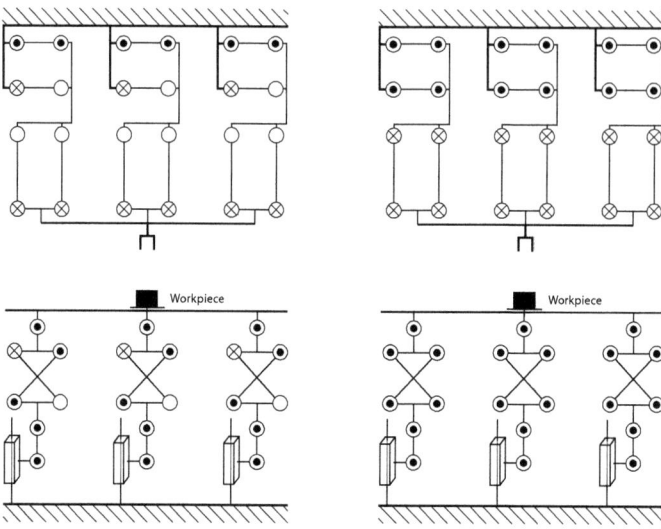

(a) Isostatic kinematic model of the MinAngle concept. The upper part (right hand) consists of a Delta kinematics.

(b) Kinematics of the realized prototype.

Figure 9.12: MinAngle kinematics.

translations, every linear movement is created using four-bar linkages, increasing the complexity again. However, the negative impact of complex kinematics is in general lowered when using flexible joints.

9.2.5 Optimization

The following list summarizes the optimization problems emerged while designing the kinematics:

- Optimizing the crossed bars in order to equally divide θ_x on their 4 pivots.

- Optimizing the geometric parameters of the whole mechanism in order to equally divide θ_y on the "top" and "bottom" pivots.

- Optimizing the placement of the TCP in order to reduce as much as possible the parasitic displacements.

9.2 MinAngle: A hybrid 5-axes High Precision Machine Tool for μEDM

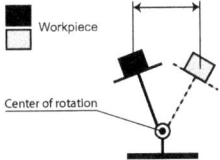
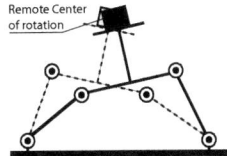

(a) Example of what happens during the rotation with a simple pivoting table. Parasitic translation is a consequence.

(b) Example of a Remote Center of Motion (RCM) mechanism. The workpiece can be placed in the center of motion and the resulting parasitic movement is very small.

Figure 9.13: Comparision between a simple rotating pivot and a RCM system

All these optimization problems were solved and are carried out in the appendix, under section A.

A negative influence of the "torsion pivot" (see figure 9.11) on the robot's stiffness has been diagnosed, while evaluating the robot's static performance. In the first design studies of the MinAngle this pivot was responsible for the execution of a small torsion angle, evaluated around $\pm 2.5°$. Although this angle is very small, and the pivot could be designed accordingly robust, its stiffness was greatly affecting the overall stiffness of the robot. This is why further investigations on this subject were made.

A great reduction of rotation (torsion) amplitude has been measured, depending on the orientation of the pivots in crossed bars (42). By optimizing the orientation of the joint axes, it is possible to minimize the torsion in order to completely eliminate the pivot. This considerably simplifies the manufacturing and enhances the stiffness of the whole system. The torsion angle could be reduced below $\pm 0.2°$. This small amount of torsion is now absorbed by a localized, elastic deformation in the kinematic chain (43) (crossed bars and their pivots). Figure 9.14 illustrates the evolution of the kinematic chains.

9. CASE STUDIES AND PROTOTYPES

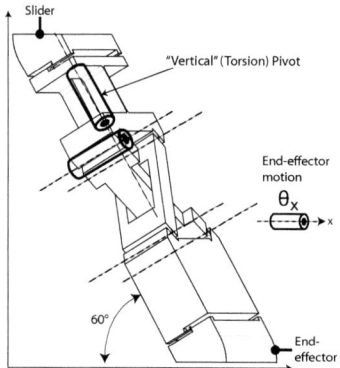
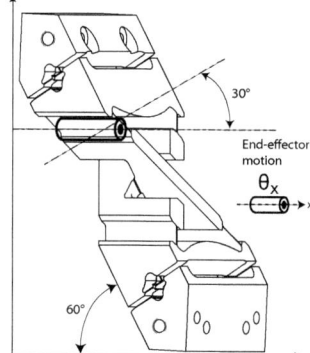

(a) First version of the kinematic chain. The pivots (of the crossed bars) were oriented in order to be perpendicular to the part's main plane, thereby simplifying the machining of this complex piece. No pivot is aligned with the illustrated output motion. A movement of all joints is needed to generate the output angle θ_x.

(b) Realized version of the kinematic chain. The pivots of the crossed bars are oriented in order to be aligned to the output angle θ_x. This way the output angle θ_x can be generated principally by the 4 pivots of the crossed bars. Torsion is considerably reduced and the vertical pivot can be eliminated.

Figure 9.14: Evolution of the pivots in the crossed bars.

9.2 MinAngle: A hybrid 5-axes High Precision Machine Tool for μEDM

The recorded parasitic displacement of the robot's output is illustrated in figure 9.15, right side. Comparisons have shown an exact accordance between measurements and simulations (43).

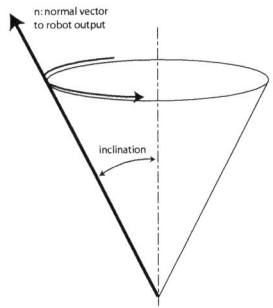
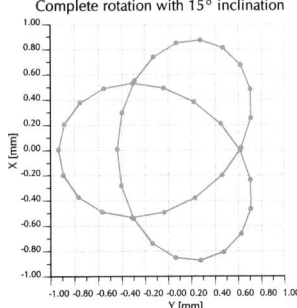

(a) Definition of the reference trajectory that was used to determine the maximum parasitic movements. The normal vector of the robot executes a trajectory on the limit of a cone with 15° half cone angle (maximum requested angular stroke).

(b) Simulated parasitic $X - Y$ displacement recorded during the reference trajectory. The maximum displacement with respect to the center is less than 1mm.

Figure 9.15: Parasitic displacement of the MinAngle's left hand.

The division of the output angle, a key feature of the MinAngle, has been verified on a CAD model. The simulations made on the virtual CAD model show a very precise division of the output angle. The results are represented in figure 9.16.

143

9. CASE STUDIES AND PROTOTYPES

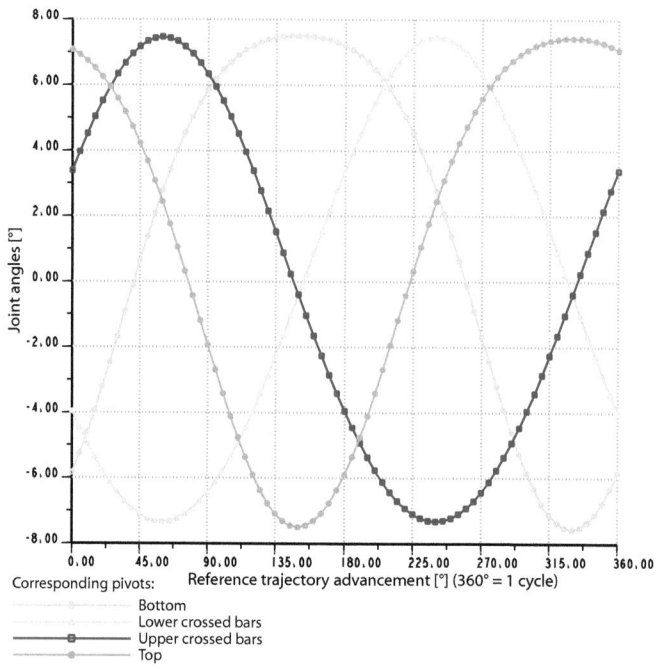

Figure 9.16: The simulation shows the evolution of all joint angles (in one kinematic chain) for the execution of the reference trajectory (see figure 9.15, left side). No joint exceeds ±7.5°, half of the output angle.

9.2 MinAngle: A hybrid 5-axes High Precision Machine Tool for μEDM

9.2.6 Technology

The decision about implementing flexure-based joints has been taken at the very beginning of the project, this mainly because of the targeted repeatability. The "Top" and "Base" pivots were designed as **separated, crossed blades pivots** (see figure 9.17). This enhances their stiffness and delays the limits of buckling (36). The robot also gains robustness against external forces or impacts.

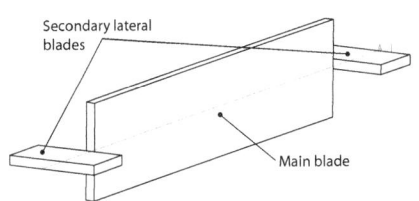

(a) Schematic illustration of the blades' arrangement. The central main blade is oriented towards the principal direction of the loads. The secondary lateral blades delay the buckling and enhance the stiffness when the pivot executes an angle. Their presence modifies the deformation mode of the main blade.

(b) Picture of the "top" pivot.

Figure 9.17: Pivot with separated, crossed blades.

The guiding and actuation of the sliders is carried out by rolling linear bearings combined with precision ball-screws. Their positioning repeatability was measured around $\pm 0.3 \mu m$. Even if these sliders are relatively precise, they proved to be the weak point in the robot.

9.2.7 Final Design Considerations

The MinAngle's left hand is equipped with a simple interfacing system which enables a fast and precise changing of the support. The left side of figure 9.18 shows a close-up on the kinematics with the actual support, which is designed for the characterization of the robot. This conical part is referenced by a precise bore and cleated on the upper reference surface of the end-effector. The precise ball – used in precision bearings – serves as the target for the measuring cube presented in section 7.6.

The conical support was replaced by a polygonal mirror, combined with an autocollimator, in order to characterize the angular repeatability of the robot. This setup is

9. CASE STUDIES AND PROTOTYPES

illustrated on the right side of figure 9.18.

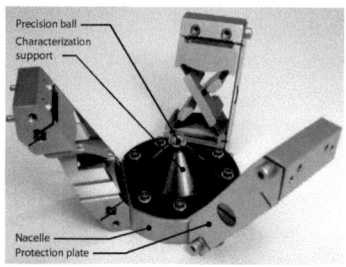 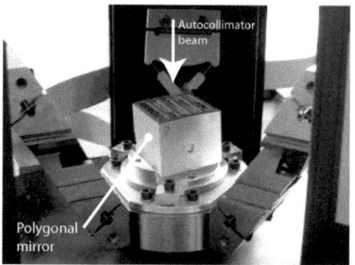

(a) A picture of the flexure-based part of the kinematics. If the mechanism is dismantled, or during transport, each arms gets a plate mounted on a side, in order to protect the joints from impacts and loads.

(b) Picture of the measurement setup used to determine the angular repeatability of the left hand.

Figure 9.18: Detail pictures of the MinAngle's left hand prototype.

9.2.8 Results and Conclusion

The table 9.4 resumes the characteristics of the MinAngle's left hand.

Table 9.4: Summary of the characteristics measured on the MinAngle's left hand.

Characteristic:	Value:
Angular stroke (θ_x/θ_y)	$\pm 15°$
Linear stroke (z)	± 5mm
Linear repeatability	$\pm 0.3\ \mu m$
1st Eigenfrequency	450 Hz
Angular Repeatability	± 0.6 arcsec
Maximum parasitic displacement	1mm
Stiffness@Output $(x/y/z)$	$3/3/9.5\ \frac{N}{\mu m}$
Machine size	$\varnothing 320$mm x 284mm

A great part of the deficiencies of repeatability and stiffness can be traced back to the linear sliders. Their own linear repeatability and stiffness is the major contributor

9.2 MinAngle: A hybrid 5-axes High Precision Machine Tool for μEDM

to the values measured at the output. In fact, because of the high vertical movement of the sliders they were realized with rolling linear bearings, ballscrew drives and encoders which were all combined in a linear module[1]. Replacing the rolling sliders by flexure-based linear guidings, optical scales and direct-drive motors would be an effective enhancement and would push the performance of the robot even higher (43).

The measurements of the 1st Eigenfrequency (evaluated at 430Hz (42), and measured around 450Hz) provide satisfying results if compared to other high-precision machines: The DeltaCube presents, according to Bacher (7) and depending on its version, Eigenfrequencies going from 200Hz (version 1) to 600Hz (version 2).
The whole MinAngle concept, both hands combined, presents a complete and space-saving 5-axes micro-machine which can be applied to several different applications. Figure 9.19 proposes a concept of a milling machining center.

The splitting of the 5 DOF, on 2 distinct hands with decoupled translations and rotations, provides a relatively vast working volume without compromising the static characteristics. Compared to the very efficient Sigma6 (35) the MinAngle provides a larger working volume with comparable stiffness values (19). The MinAngle's left hand proved to be a good complement for the well established Delta. The kinematics might not be the most effective, however, they have the capacity of reaching high angles without affecting the stiffness.

---

[1]The easiness of implementing the complete, commercialized linear module for a very first prototype was a reason too.

9. CASE STUDIES AND PROTOTYPES

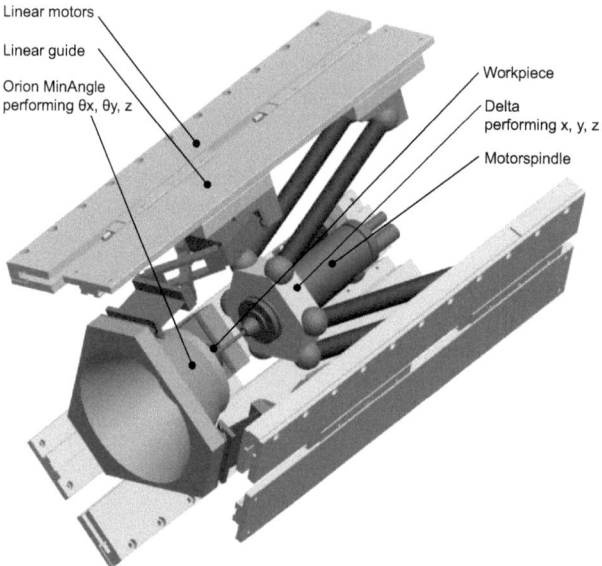

Figure 9.19: Example of the MinAngle concept applied to a different application: High-speed machining. The concept proposes to use direct drive linear drives. With this combination of left and right hand it is possible to use the same linear guidings and magnet rails for both hands.

9.3 Omikron5: A hybrid 5-axes Machine Tool

9.3.1 Introduction

The Omikron5 concept has been developed in collaboration with *Mikron Machining Technology*[1] to carry out the deburring of small workpieces with a high dexterity. The goal was to create a 5-axes module which can be integrated in an existing machining center without any modification of the latter. The deburring of these thin features, which was up to now accomplished on another machine, should be integrated in order to reduce the amount of machines and setups for a complete finalization. "One setup for an out-and-out finalized piece" was the maxim of this development. The main challenges were the integration in an existing machine, with very few available place, and the required angles of rotation. The machine had to provide a very high ratio of working volume size over footprint size.

As already insinuated before, the projected machine had to be designed as an autonomous and complete module, integrable at different places in the existing machining center.

The realized prototype is represented in figure 9.20.

9.3.2 Project Requirements

Table 9.5: Summary of the Omikron5 project requirements.

Characteristic:	Value:
Needed mobility	5 (3T2R)
Linear stroke (x/y/z)	±35mm
Angular stroke (θ_x/θ_y)	±90° / ±90° (5-side machining)
Acceleration	$50\frac{m}{s^2}$ (5g)
Precision	0.05mm
Power of spindle	150 W
Modularity aspects	Multiple modules must work on same workpiece.
Integration aspects	Mountable on an existing machine, without any modification, and, at different places in the machine.

---

[1] www.mikron.com

9. CASE STUDIES AND PROTOTYPES

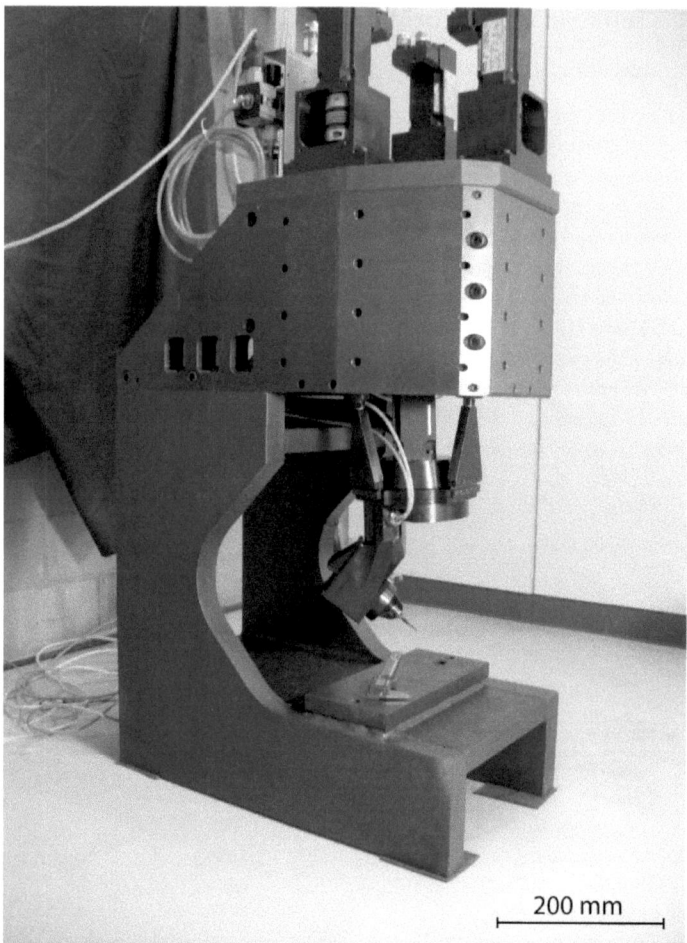

Figure 9.20: Omikron5: A 5-axes machine tool for high-speed 5-side machining. The L-shaped support on the lower half of the picture was designed to maintain the module as long as it is not integrated in the existing machine.

9.3 Omikron5: A hybrid 5-axes Machine Tool

9.3.3 Distribution of Mobilities

The project requirements completely limited the choice of distributing the mobilities. In fact, as there can be several modules machining at the same time, the workpiece cannot be moved. A moving workpiece, whose movement is interpolated with the 1st module, would force the 2nd module to be inactive. These restrictions forced us to place all DOF on the right hand. Figure 9.21 illustrates the possibility of arranging multiple modules around the same workpiece.

Furthermore, regarding the high angle amplitudes and the technical feasibility, the possibility of integrating a fully-parallel mechanism dropped. The rotations imperatively had to be based on a serial-kinematic wrist.

Figure 9.21: Example of 2 modules working on the same setup. The way the 2 modules are arranged allows to cover the complete angular range (complete sphere). The cube in the center represents the working volume of the modules.

9. CASE STUDIES AND PROTOTYPES

9.3.4 Kinematics

From the beginning on the kinematics were separated in two parts – according to the type of movement – the translator and the double-tilting head. Because of the rotation amplitudes the head had to be based on a serial kinematics, whereas the translator, because of the targeted accelerations, had to be based on a parallel kinematics.

The design of the existing machine – where the module has to be built in – imposes a restricted available space, and both parts of the kinematics have to consider this.

Translator

Many different parallel-kinematic translators were investigated, including the well-performing Delta. However, because of the limited space, another concept was needed. In order to execute large translations in a very limited space a special concept was used: The translation is generated by a rotation, converted in a linear displacement by means of a lever. Two kinematics were chosen and analyzed in detail, they are illustrated in figure 9.10 (in section 9.2). Both of these kinematics are usually known and used as tilting platforms. One of these kinematics proved to be very efficient for the above-mentioned space problem, the Dunlop kinematics. Figure 9.22 illustrates where and how the movement is generated thanks to this kinematics. Figure 9.23, right side, gives an inside view of the mechanics of the translator.

The problem that is created because of our concept is obvious: The translation is generated, but, there is also a small amount of rotation that takes place. The translator therefore generates a non-pure movement. These rotations therefore need to be taken into account and *compensated by the tilting head*.

Double-tilting Head

The head has to execute rotations without introducing parasitic translation[1]. Therefore, a kinematics with RCM capability is used. Both rotation axes are arranged in order to intersect in the TCP. The head is illustrated in figure 9.23.

The complete kinematics of the Omikron5, translator and head assembled, can be examined in figure 9.24. The mobility inefficiency of the Omikron5 amounts 3.4 (total mobilities of 17, and 5 effective mobilities of the end-effector). The kinematics proves to be very effective. The amount of joints is low, therefore considerable limiting the risks. In fact, only the 3 spherical joints are novel and therefore critical, all other elements

---
[1] Again, because of the limited space.

9.3 Omikron5: A hybrid 5-axes Machine Tool

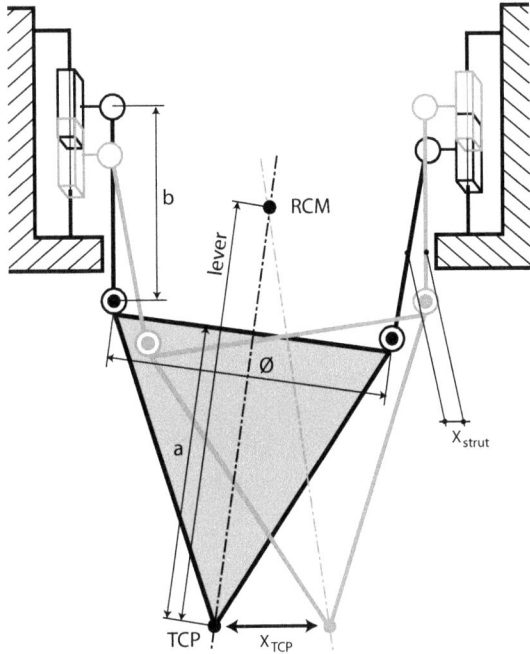

Figure 9.22: The translation of the TCP is created by a rotation of the parallel-kinematic part. This rotation takes place around a virtual rotation point, by this, limiting the displacements in the upper machine part. The effective displacement of the TCP X_{TCP} is much bigger than the displacements induced in the upper machine part X_{strut}. A pure translational mechanism (e.g. the Delta) could not create that much translation in this restricted space.

are well-known. From an industrial point of view the kinematics of the Omikron5, although being hybrid, presents a minimal risk.

9.3.5 Optimization

The restricted space available did not give a lot of clearance for optimization. In fact, the rather thin, but long, available space forced us to capitalize the whole width of the space, choosing therefore the parameter ∅ (see figure 9.22) as big as possible.

9. CASE STUDIES AND PROTOTYPES

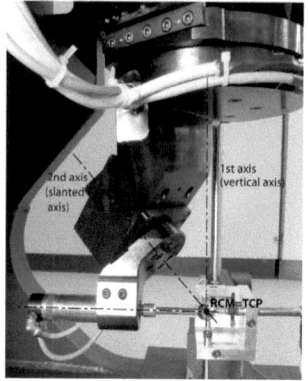 

(a) Omikron5's RCM head with the definition of the axes. The tool can be tilted till it reaches the horizontal plane.

(b) The very restricted space requires a compact upper machine design.

Figure 9.23: Pictures of the tilting head and translator of the Omikron5.

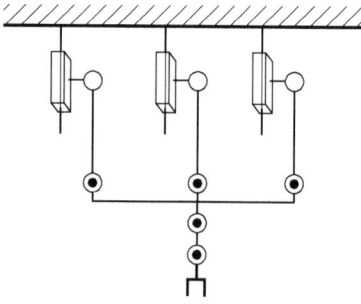

Figure 9.24: Isostatic kinematic model of the Omikron5.

This was done in order to get the ratio $\frac{\varnothing}{a}$ as big as possible. This ratio determines the "transmission ratio" between the vertical movement of the slider (articular coordinates) and the generated translation. The higher the ratio, the more sensitive the translation is, and in addition, the stiffer the platform is.

9.3 Omikron5: A hybrid 5-axes Machine Tool

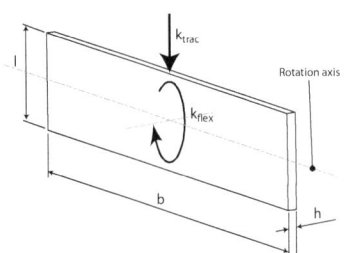

(a) A flexible pivot of the Omikron5. The bending length l is perfectly controlled by the clamping/fixation of the blade.

(b) Definition of geometric parameters and stiffness values.

Figure 9.25: Omikron5's flexible pivots. Parameterization and realization.

The parameter **a** was reduced as much as possible until it reached the minimum size for the implementation of the spindle. The parameter **b** resulted from the choice made for **a**. In fact, the sum (**a** + **b**) has a minimal length allowing the machine to reach the lowest points of the working volume.

The second variable to optimize were the angles of the pivots. *This was an optimization task issued from a technological choice.* The goal was to reduce the angles of the pivots which would allow the integration of flexible pivots. A great reduction of angle has been achieved by a small modification of the kinematics[1], changing the kinematics of the translator from a Orion (figure 9.10, left side) to a Dunlop (figure 9.10, right side). Marti (46) measured a reduction from ±6° to ±1.7°, which finally enabled a more robust design of the pivot.

9.3.6 Technology

As mentioned before, *the pivots* of the parallel-kinematic part were realized using flexible blades, allowing them to get very high stiffness values. The dimensions of the pivots are represented in figure 9.25. The stiffness values were calculated according to Henein (36), and presented in table 9.6.

The *linear sliders* were realized using conventional linear bearings and ballscrew drives. This has been decided in order to use the same standard elements which are

---
[1] The changes are taken into account in figure 9.24. The paragraph just illustrates a design iteration that precedes the final design.

9. CASE STUDIES AND PROTOTYPES

Table 9.6: Characteristics of flexible pivot

Characteristic:	Value:
Stiffness k_{trac}	$1930 \frac{N}{\mu m}$
Stiffness k_{flex}	$1 \cdot 10^6 \frac{Nm}{rad}$
h	$500 \mu m$
b	80mm
l	4mm
Achievable angle	$\pm 3.5°$

already present in the existing machining center. Figure 9.26 provides a detailed view on a slider.

The *spherical joints* are an industrialized and simplified version of the 3rd evolution spherical joint presented in section 6.2. The following modifications were made:

The preloading system has been designed for a preload of 1200 N and its pins can be locked to avoid dismantling.

The coupling membrane allowing a switching between spherical- or universal-joint mode has been removed. The joint needs all 3 DOF.

Protection scrapers were introduced to keep the interfaces clean.

The *rotation axes* of the serial-kinematic tilting head were realized using complete harmonic drive units. Both axes possess a single, crossed-roller bearing in order to guide the axes. To guarantee a high stiffness, the diameters of the bearings were chosen as big as possible (with respect to the available space).

1st axis/Vertical axis: This axis was realized using a harmonic drive CPU-M-25 unit combined with a conventional servomotor. The unit contains the output bearing and a harmonic drive gearbox with a reduction ratio of 50. Its tilting stiffness comes to $3.9 \cdot 10^5 \frac{Nm}{rad}$ and its torsional stiffness to $2.4 \cdot 10^4 \frac{Nm}{rad}$.

2nd axis/Slanted axis: This axis, because of the very limited space at the end of the robot, was realized using a complete harmonic drive FHA-C-11 Mini motor-unit. This motor-unit contains all necessary elements: Motor, harmonic drive gearbox (with a reduction ratio of 30), position sensor (encoder) and the output bearing. Its tilting stiffness comes to $4 \cdot 10^4 \frac{Nm}{rad}$ and its torsional stiffness to $0.84 \cdot 10^3 \frac{Nm}{rad}$.

9.3 Omikron5: A hybrid 5-axes Machine Tool

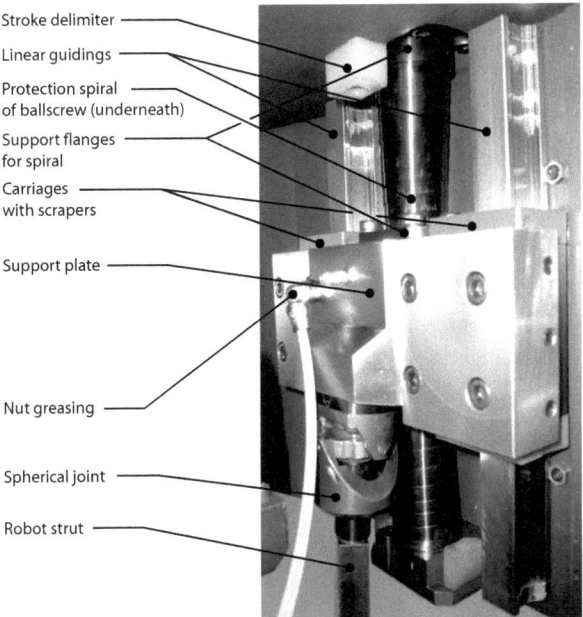

Figure 9.26: Detailed view of the slider. The ballscrew is hidden by the protection spirals, the nut is integrated in the support plate.

This motor-unit is illustrated in figure 9.27. Its compactness and performances made it a very vital element of the module.

Both reduction ratios were chosen in order to be close to the theoretical optimum (best possible acceleration, see section 3.1.3).

9.3.7 Final Design Considerations

As the module is designed for machining, a very special attention to protection issues was paid. Considering the high mobility of the machine – in terms of DOF and amplitudes – and the big swiveling radius of the spindle we decided, in a first approach,

9. CASE STUDIES AND PROTOTYPES

Figure 9.27: Harmonic drive module FHA-C Mini from Harmonic Drive AG (www.harmonicdrive.de). The module contains motor, harmonic drive gearbox, position sensor and output bearing. The module fits in a cube of 60mm side length.

for an *element-wise protection*[1]. Every element had to be protected except the flexible pivots.
Figure 9.28 shows a cut through the Omikron5 and highlights the protection elements of the 2 pivots, the tilting head, and of the spherical joints. Figure 9.26 shows the protection elements introduced for the sliders and their ballscrews.

The air sealing and cabling of the spindle were conducted through the different parts of the tilting head and through the hollow shaft of the harmonic drive. This way, the cables and the air sealing tube are not apparent and won't get damaged or squeezed. Their passage through the mechanism is highlighted in figure 9.28.

9.3.8 Results and Conclusion

The table 9.7 resumes the characteristics of the Omikron5.

The hybrid kinematics proved to be completely adapted (and necessary) to the very restricted available space and satisfies the required working volume. Furthermore, the module does not need any modification on the existing machining center and can be mounted on the 2 different, intended slots.
The repeatability measurements prove the high quality of all joint designs. The repeatability of the Omikron5 lies at $1\mu m$. This generally, after calibration, leads to a

---
[1]See section 7.4 for more information.

9.3 Omikron5: A hybrid 5-axes Machine Tool

Table 9.7: Summary of the Omikron5's measured characteristics.

Characteristic:	Value:
1st Eigenfrequency	80 Hz
Acceleration	n.a. (intended drives not yet implemented)
Repeatability (x/y/z)	$\pm 0.88 \mu m / \pm 0.88 \mu m / \pm 0.23 \mu m$
Angular stroke (θ_x/θ_y)	$\pm 90° / \pm 90°$
Linear stroke (x/y/z)	$\pm 35 mm / \pm 35 mm / \pm 50 mm$
Stiffness (x/y/z)	2.5-3/2.5-3/10 $\frac{N}{\mu m}$

precision which is lower than the required $50 \mu m$. Furthermore, the measured stiffness values are by far sufficient for deburring tasks (63). The slightly lower eigenfrequency (intended at 100Hz) can be due to several points: quality of contact between fastened parts, flexure pivot's stiffness while bent, varying preload in spherical joints.

In general, the low complexity of the kinematics and the general mechanical performance of this first prototype are very satisfying. The ongoing and future improvements will push the performance even higher.

9. CASE STUDIES AND PROTOTYPES

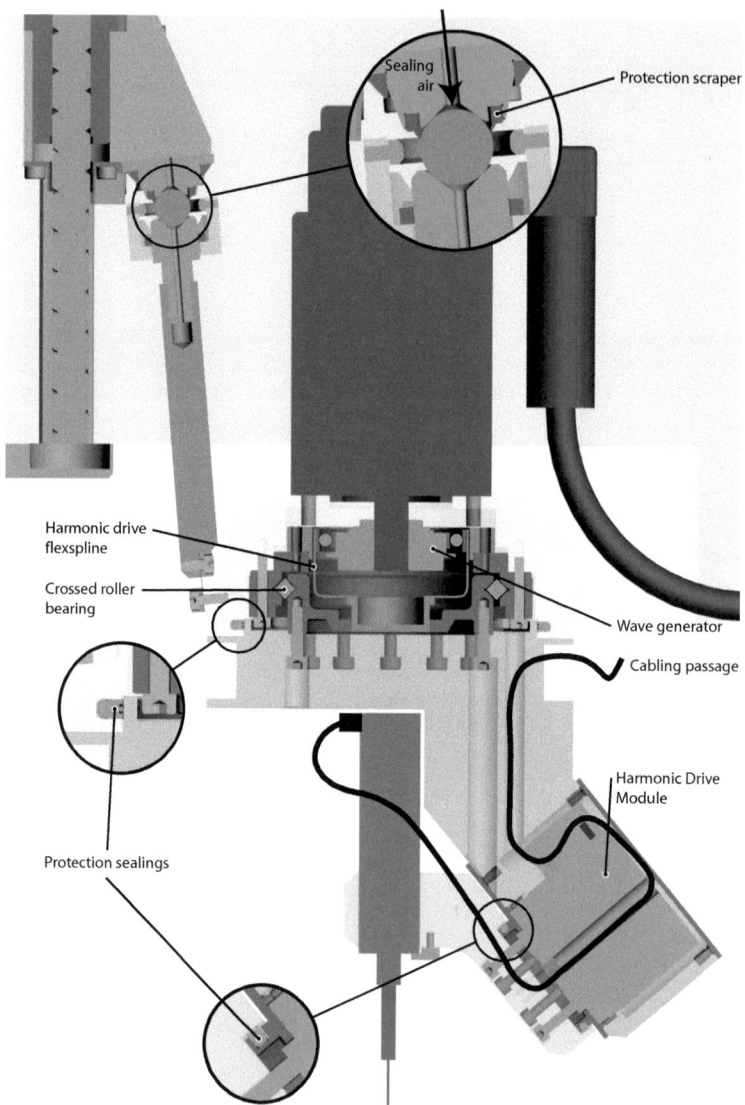

Figure 9.28: Detailed cut through the Omikron5. The single protection elements as well as the cable passage of the tilting head are shown.

10
Conclusion

10.1 Follow-up of the Results

This thesis resumes the work done, and the results obtained, in multiple projects in the domain of machine tools. All projects were conducted as collaboration with industrial partners and aimed for a future commercialization of the resulting prototypes. In the scientific domain, the work places itself in a series of theses about the design of high-performance parallel/hybrid mechanisms, a subject that has a long tradition in the *LSRO Laboratoire de Systèmes Robotiques*.

The originality lies in the use of hybrid kinematics for machine tools, which constitute a very effective trade-off when it comes to their viability in industrial environments. For 5-axes mechanisms and 5-side machining their lowered complexity reduces the amount of elements (joints, protections) and therefore the risks of a machine break-down. Their performance, in terms of working volume, approaches the performance of standard machines while still having partially the stiffness-, precision- and dynamical potential of fully-parallel mechanisms. A trade-off which is appreciated by the industry, as they are often sceptic about the very wide technological change (and the resulting risks) towards fully-parallel mechanisms.

In general, hybrid kinematics offer possibilities in the arrangement of axes, between a parallel or serial arrangement and between left and right hands, which gives the designer the chance to adapt the axes in a specific and selective way to their targeted performances.

The realized prototypes present very good mechanical properties and fulfill their respective project requirements, even though a multitude of compromises have been made about their geometric parameters. In fact, the integration of the mechanisms into existing machines limited the choices when tuning the geometric parameters. Here again

10. CONCLUSION

the high flexibility of a hybrid concept was helpful.

10.2 Contributions

The contributions of this work can be separated in different domains:

10.2.1 Joint Family

- *A family of high-performance spherical joints was developed*. The last evolution presents a simplified and moderately priced design. This could be guaranteed by using a precision bearing ball as referencing element of the joint, thus also guaranteeing a high precision. Furthermore, a special preloading system was developed in the form of an outer universal joint. This system allows an *uncomplicated switch from a 3-DOF spherical joint to a 2-DOF universal joint* using the same joint design and therefore keeping good mechanical characteristics.

- *A stiffness modeling for the different contact types* was deduced from the Hertzian contact theory.

- *A simple and effective process to enhance the stiffness* of gliding spherical joints was developed and successfully tested.

10.2.2 New Kinematics, Mechanisms and Prototypes

- *A catalogue of new kinematics reaching from 2 to 5 axes*, intended to be used as modules or complete kinematics.

- An increased complexity of fully-parallel mechanisms with high rotation strokes could be shown, and *the efficiency of hybrid mechanisms* for these requirements could be underlined. Besides, *the diversity of axis arrangements (parallel/serial) and distributions (left/right hand)*, which allows a high flexibility and ability to adapt, could be showed.

- *3 Case studies have been conducted and the resulting prototypes shown*, all of them being new kinematics, **the MinAngle concept, the Stewart3 and the Omikron5**. Their geometries were optimized and their geometric models were developed. Furthermore, *a generic characterization setup was designed* and used to validate the concepts the machines.

10.2.3 Design Methodology

- A *design methodology* which, besides explaining all the design steps from the start to the end, points out the most important design steps. For most of these steps this thesis proposes methods, catalogues or tools.

- A detailed list of *important aspects for the industrialization of such mechanisms* has been compiled, some of these aspects being often neglected by the scientific community. The presented concepts were kept simple, effective and straightforward.

Considering the different contributions of this work we can redraw figure 2.1 presented at the beginning[1]. The resulting impacts on the different aspects can examined in figure 10.1.

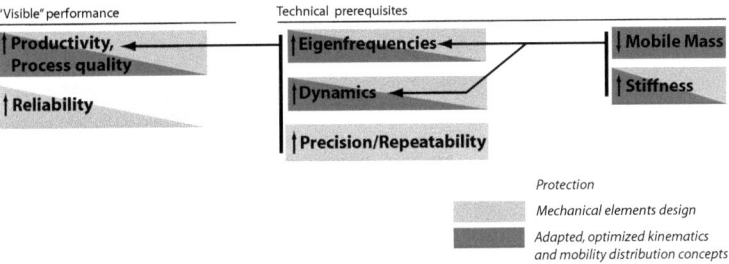

Figure 10.1: Targeted technological and economic aspects, and the contributions of this work. The subjects and contributions of this work do affect, directly or indirectly, all targeted aspects.

10.3 Limits and Perspectives

The **limits of the thesis** can be summarized as follows:

- Most of the mechanical aspects were discussed. However, impacts of the drives, the calibration, the control and their interaction with the mechanics were not investigated in details. The assumption that *"well designed mechanics present a good base for all later development"* was made.

---

[1] In the aims of the thesis, section 2.

10. CONCLUSION

- The thesis is focused on machine tools and their requirements. The prototypes, kinematics and elements were designed while keeping the required performance of machine tools in mind. For different applications (e.g. handling), most aspects have to be reconsidered and optimized with respect to different requirements.

- The concepts and methods presented in this work privilege a simple access and effectiveness over rigorous scientific methods. Instead of presenting exact and general models we prefer sensitizing the designer and attract his attention to the determining design parameters, hoping to develop a certain feeling for parallel/hybrid mechanisms.

- The realized prototypes are in their 1st version and, although they present good mechanical characteristics, there is still room for further improvements.

The **perspectives for future research:**

- *Joints:* The reliability of their design should be investigated for different materials, different preloads, surface qualities, surface treatments and lubrications. The exact shape of the plastically deformed contact region should be investigated for different stamping forces, different ball diameters[1] and the different joint materials. This, in order to find the optimum process parameters.

- *Kinematics:* Developing new kinematics, a never ending story, should be carried on. Their influence on the performance, as already mentioned, is determining.

- *Modeling:* Establishing effective, analytic, geometric models and dynamic models of the three prototypes, this, in order to provide an exact model for optimization tasks and to improve the control.

10.4 Final Note

This thesis covers the most important aspects of the realization of hybrid- and parallel-kinematic mechanisms. Several new kinematics and elements were proposed and their development through the most important steps was illustrated. They proved to be performing well and, from an industrialization point of view, proved to be effective. Hybrid kinematics can be designed to be more efficient and appropriate than fully-parallel kinematics, as they offer more possibilities in their axis arrangement. The often criticized complexity of parallel mechanisms can be reduced and the limited working

---

[1] Especially the difference of diameter between the ball used for stamping, and the ball which will be placed in the working joint.

volume be enhanced by selective use of serial axes.

We hope that this work can contribute to the inspiration of the designers, in the best case animate their creativity, and result in a popularization of these mechanisms in the industry.

10. CONCLUSION

References

Note: The pagenumbers placed after each bibliographic entry indicate on which pages the reference was cited.

[1] **Nouvelle Génération pour le leader de la machine-outil: DS Technologie introduit l'ECOSPEED sur le marché.** *Fakt DS Technologies*, 2003. ix, 40

[2] http://robotool.ifw.uni-hannover.de/pages/listeenglisch.htm, status of 2008. Robotool Website. 32, 34

[3] http://www.cecimo.be, status of 2008. CECIMO, European Committee for Cooperation of the Machine Tool Industries. 5, 6

[4] Y. ALLEMAND. **Conception et réalisation d'un mécanisme de positionnement angulaire de haute précision**, 2004. Semester Project, EPFL. 53

[5] I.I. ARTOBOLEVSKI. *Mechanisms in Modern Engineering Design.* MIR Publishers, 1977. 32

[6] M. AUBIN AND AL. *Système Mécaniques: Théorie et Dimensionnement.* Dunod, Paris, 1992. ISBN 2 10 001051 4. 70

[7] J.-P. BACHER. *Conception de robots de très haute précision à articulations flexibles: Interaction dynamique-commande.* PhD thesis, Ecole Polytechnique Fédérale de Lausanne EPFL, 2003. 51, 147

[8] J.-P. BACHER, S. BOTTINELLI, J.-M. BREGUET, AND R. CLAVEL. **Delta3: A new Ultra-high Precisions Micro-Robot, Design and Control of a Flexure Mechanism.** *Journal Européen des Systèmes Automatisés*, Volume 36, No. 9, pages 1263–1275, 2002. 137

[9] I. BONEV. http://www.parallemic.org, status of 2008. ParalleMIC – The Parallel Mechanisms Information Center. 32

[10] R. BOUDREAU AND C.M. GOSSELIN. **The synthesis of planar parallel manipulator with a genetic algorithm.** *ASME Journal of Mechanical Design*, Volume 121, Issue 4, pages 533–537, 1999. 105

REFERENCES

[11] T. BROGARDH. **PKM Research – Important Issues, as seen from a Product Development Perspective at ABB Robotics**. *Proceedings on Fundamental Issues and Future Research Directions for Parallel Mechanisms and Manipulators*, 2002. Quebec City, Canada. 32, 34, 66

[12] J. R. CANNON. **Compliant mechanisms to perform bearing and spring functions in high precision applications**, 2004. Master Thesis, Brigham Young University, available at http://contentdm.lib.byu.edu/ETD/image/etd599.pdf (september 2008). 53, 54

[13] D. CHABLAT. **Device for the movement and orientation of an object in space and use thereof in rapid machining**, International Patent No. PCT/WO/2004/071705, 2006. 26, 57, 59, 207

[14] R. CLAVEL. **Dispositif pour le déplacement et le positionnement d'un élément dans l'espace**, Swiss Patent No. 672089 A5, 1985. 10, 43, 44

[15] R. CLAVEL. **DELTA, a fast robot with parallel geometry**. *Proceedings of the ISIR International Symposium on Industrial Robots*, 1988. Lausanne, Switzerland. 10, 43

[16] R. CLAVEL. **Composants de la Microtechnique**. Réprographie EPFL, Lausanne, 2003. 72

[17] R. CLAVEL AND H. BLEULER. *Robotique, Microrobotique*. Réprographie EPFL, Lausanne, 2004. 22

[18] R. CLAVEL, M. BOURI, S. GROUSSET, AND M. THURNEYSEN. **A new 4 DOF parallel robot: The Manta**. *Keynote speech, Proceedings of the international workshop on parallel kinematics machines PKM99*, 1999. Milano, Italy. 59, 207

[19] R. CLAVEL, P. PHAM, B. LORENT, B. LE GALL, AND M. BOURI. **New Variants of Delta Robots and Double-Tilt Platform for Assembly**. *Proceedings of the 3rd International Colloquium of the Collaborative Research Center*, 2008. Braunschweig, Germany. 31, 147

[20] R. CLAVEL, M. THURNEYSEN, J. GIOVANOLA, AND SCHNYDER M. **A New Parallel Kinematics Able to Machine 5 Sides of a Cube-Shaped Object – Hita STT**. *Proceedings of the First International Colloquium on Robotic Systems for Handling and Assembly*, 2002. Braunschweig, Germany. 46

[21] V. COLLADO AND S. HERRANZ. **Space 5H - A New Machine Concept for 5-Axis Milling of Aeronautic Structural Components**. *Proceedings of the PKS Chemnitz Parallel Kinematic Seminar*, 2004. Chemnitz, Germany. 4, 41, 63

[22] H. CZICHOS (EDITOR). *Hütte: Die Grundlagen der Ingenieurwissenschaften*. Springer, Berlin, 1991. ISBN 3-540-19077-5. 70

[23] T. CZWIELONG AND W. ZARSKE. **PEGASUS – Incorporating PKM into Woodworking**. *Proceedings of the PKS Chemnitz Parallel Kinematic Seminar*, 2002. Chemnitz, Germany. 42, 63

REFERENCES

[24] M.-O. DEMAUREX. *Approche théorique de la conception de la structure mécanique d'un robot industriel.* PhD thesis, Ecole Polytechnique Fédérale de Lausanne EPFL, 1979. 23

[25] G. R. DUNLOP AND T. P. JONES. **Position analysis of a 3-DOF parallel manipulator.** *Mechanism and Machine Theory, Volume 32, No. 8, pp.903-920,* 1997. 137

[26] H. FRAYSSINET. *Méthode d'étalonnage d'une machine-outil à cinématique parallèle à cinq axes à grands angles d'inclinaison.* PhD thesis, Ecole Polytechnique Fédérale de Lausanne EPFL, 2007. 105

[27] F. GAO, B. PENG, H. ZHAO, AND W. LI. **A novel 5-DOF fully parallel kinematic machine tool.** *International Journal of Advanced Manufacturing Technologies, Volume 31,* pages 201–207, 2006. 59, 208

[28] K.+R. GIECK. *Formulaire Technique.* Gieck Verlag, Germering, Germany, 1997. 10ème edition française, ISBN 3 920 379 241. 86, 87

[29] T. GMÜR, M. DEL PEDRO, AND J. BOTSIS. *Mécaniques des structures.* Réprographie EPFL, Lausanne, 1999. 87

[30] L. E. GOODMAN AND L. M. KEER. **The Contact Stress Problem for an Elastic Sphere Indenting an Elastic Cavity.** *International Journal of Solids and Structures, Volume 1,* pages 407–415, 1965. 71

[31] C. GOSSELIN, E. ST. PIERRE, AND M. GAGNÉ. **On the Development of the Agile Eye.** *IEEE Robotics and Automation Magazine, Volume 3, Issue 4,* pages 29–36, 1996. 94

[32] H. GRONBACH. **TriCenter – A Universal Milling Machine with Hybrid Kinematic.** *Proceedings of the PKS Chemnitz Parallel Kinematic Seminar,* 2002. Chemnitz, Germany. 63

[33] M. GRUEBLER. *Getriebelehre – Eine Theorie des Zwanglaufs und der ebenen Mechanismen.* Springer Verlag, Germany, 1917. ISBN 1 429 742 720. 7

[34] L. C. HALE. *Principles and techniques for designing precision mechanisms.* PhD thesis, Massachusetts Institute of Technology MIT, 2006. 32, 33, 61

[35] P. HELMER. *Conception systématique de structures cinématiques orthogonales pour la microrobotique.* PhD thesis, Ecole Polytechnique Fédérale de Lausanne EPFL, 2006. 31, 51, 66, 135, 147

[36] S. HENEIN. *Conception des structures articulées à guidages flexibles de haute précision.* PhD thesis, Ecole Polytechnique Fédérale de Lausanne EPFL, 2000. 50, 52, 68, 135, 137, 145, 155, 181, 182

[37] N. HENNES. **ECOSPEED - An Innovative Machinery Concept for High-Performance 5-Axis-Machining of Large Structural Components in Aircraft Engineering.** *Proceedings of the PKS Chemnitz Parallel Kinematic Seminar,* 2002. Chemnitz, Germany. ix, 3, 4, 40, 63

REFERENCES

[38] N. HENNES AND D. STAIMER. **Application of PKM in Aerospace Manufacturing – High Performance Machining Center ECOSPEED, ECOSPEED-F and ECO-LINER**. *Proceedings of the PKS Chemnitz Parallel Kinematic Seminar*, 2004. Chemnitz, Germany. 40

[39] J. HESSELBACH, J. RAATZ, J. WREGE, AND S. SOETEBIER. **Design and Analyses of a macro parallel robot with flexure hinges for micro assembly tasks**. *Proceedings of the ISR International Symposium on Robotics*, 2004. Paris, France. 139

[40] C. JOSEPH. *Contribution à l'accroissement des performances du processus de microEDM par l'utilisation d'un robot à dynamique élevée et de haute précision*. PhD thesis, Ecole Polytechnique Fédérale de Lausanne EPFL, 2005. 137

[41] D. KANAAN, PH. WENGER, AND D. CHABLAT. **Kinematics analysis of the parallel module of the VERNE machine**. *Proceedings of the 12th IFToMM World Congress*, June 18-21, 2007. Besançon, France. 41

[42] P. KOBEL. *La conception d'un système de positionnement angulaire entièrement flexible et à grande course pour l'intégration dans une machine-outil*, 2006. Semester Project, EPFL. 141, 147

[43] P. KOBEL. *Caractérisation en optimisation d'une machine parallèle 3-axes d'ultra-haute précision à articulations flexibles: concept MinAngle*, 2007. Master Thesis, EPFL. 116, 141, 143, 147

[44] S. KRUT. *Contribution à l'étude des robots parallèles légers, 3T-1R et 3T-2R, à fort débattement angulaires*. PhD thesis, Université de Montpellier II, 2002. 26, 57

[45] R. KUSTER. *Réalisation et commande d'un nouveau robot d'assemblage micro-Delta linéaire*, June 2004. Semester Project, EPFL. 83

[46] J. MARTI. *Design of a hybrid kinematics high-dynamic machine tool*, 2007. Master Thesis, EPFL. 155

[47] Y. S. MARTIN, M. GIMÉNEZ, M. RAUCH, AND J.-Y. HASCOËT. **Verne – A New 5-Axes Hybrid Architecture Machining Centre**. *Proceedings of the PKS Chemnitz Parallel Kinematic Seminar*, 2006. Chemnitz, Germany. 41, 63

[48] J.-M. MERLET. *Parallel Robots*. Kluwer Academic Publishers, 2001. ISBN 1-4020-0385-4. 97, 100, 102

[49] J.-P. MERLET. **Still a long way to go on the road for parallel kinematics**. *Proceedings of the ASME 2002 Design Engineering Technical Conference DETC Conference*, 2002. Montreal, Canada. 31

[50] J.-P. MERLET. **Dimensional Synthesis of Parallel Robots with a Guaranteed Given Accuracy over a Specific Workspace**. *Proceedings of the ICRA 2005 International Conference on Robotics and Automation*, 2005. Barcelona, Spain. 102

REFERENCES

[51] J.-P. MERLET. http://www-sop.inria.fr/coprin/equipe/merlet, status of 2008. Jean-Pierre Merlet's Website. 32

[52] V. NABAT. *Robots parallèles à nacelle articulée: Du concept à la solution industrielle pour le pick-and-place.* PhD thesis, Université Montpellier II, 2007. 26, 44, 45

[53] V. NABAT, F. PIERROT, AND AL. **High-speed parallel robot with four degrees of freedom**, International Patent No. PCT/WO/2006/087399, 2006. 57

[54] R. NEUGEBAUER. **Parallel Kinematic Structures in Manufacturing**. *Proceedings of the PKS Chemnitz Parallel Kinematic Seminar*, 2002. Chemnitz, Germany. ix, 3

[55] K.-E. NEUMANN. **Tricept Robot US Patent 4'732'525**, 1988. Neos Product HB, Sweden. 12

[56] K.-E. NEUMANN. **Exechon Concept**. *Proceedings of the PKS Chemnitz Parallel Kinematic Seminar*, 2006. Chemnitz, Germany. 39, 61, 63

[57] E. PERNETTE. *Robot de haute précision à 6 degrés-de-liberté pour l'assemblage des microsystèmes.* PhD thesis, Ecole Polytechnique Fédérale de Lausanne EPFL, 1998. 51, 52, 137

[58] P. PHAM. **Alpha5: Commande d'axes par bus de Terrain Profibus**, 2003. Master Thesis, EPFL. 27

[59] P. PHAM, M. BOURI, M. THURNEYSEN, AND R. CLAVEL. **Profibus PC based Motion Control with Application to a new 5 Axes Parallel Kinematics**. *Proceedings of the ISR International Symposium on Robotics*, 2004. Paris, France. 27, 28, 57

[60] P. PHAM AND B. BRÜHWILER. **Alpha5: Un robot parallèle à 5 DDL**, 2002. Semester Project, EPFL. 26, 27

[61] P. PHAM, Y.-J. REGAMEY, M. FRACHEBOUD, AND R. CLAVEL. **Orion MinAngle: A flexure-based, double-tilting parallel-kinematics for ultra-high precision applications requiring high angles of rotation**. *Proceedings of the ISR International Symposium on Robotics*, 2005. Tokyo, Japan. 68, 135, 139

[62] D. PRUST. **High Performance Machining Center VISION**. *Proceedings of the PKS Chemnitz Parallel Kinematic Seminar*, 2004. Chemnitz, Germany. 35

[63] F. PRUVOT. *Conception et Calcul des Machines-Outils.* Presses Polytechniques et Universitaires Romandes, Lausanne, 1993. ISBN 2-88047-248-X. 4, 5, 121, 159

[64] M. SCHNYDER, J. GIOVANOLA, R. CLAVEL, M. THURNEYSEN, AND D. JEANNERAT. **Spherical Joints with 3 and 4 Degrees of Freedom for 5-Axis Parallel Kinematics Machine Tool**. *Proceedings of the PKS Chemnitz Parallel Kinematic Seminar*, 2004. Chemnitz, Germany. 49

REFERENCES

[65] E. SCHOPPE, A. PÖNISCH, V. MAIER, T. PUCHTLER, AND S. IHLENFELDT. **Tripod Machine SKM 400 – Design, Calibration and Practical Application**. *Proceedings of the PKS Chemnitz Parallel Kinematic Seminar*, 2002. Chemnitz, Germany. 48

[66] N. SCLATER AND N. P. CHIRONIS. *Mechanisms and Mechanical Devices Sourcebook*. 3rd Edition, McGraw Hill, 2001. 32

[67] M. SCUSSAT. *Assemblage bidimensionnel de composants optiques miniatures*. PhD thesis, Ecole Polytechnique Fédérale de Lausanne EPFL, 2000. 203

[68] A. H. SLOCUM. *Precision Machine Design*. Society of Manufacturing Engineers, 1992. ISBN 0-13-690918-3. 61

[69] G. SPINNLER. *Conception des Machines – Principes et Applications*. Presses Polytechniques et Universitaires Romandes, 2002. Volume 1, ISBN 2-88074-301-X. 74

[70] M. THURNEYSEN. *Méthode systématique de conception de cinématiques parallèles*. PhD thesis, Ecole Polytechnique Fédérale de Lausanne EPFL, 2004. 46

[71] M. THURNEYSEN, R. CLAVEL, M. BOURI, H. FRAYSSINET, GIOVANOLA J., M. SCHNYDER, AND D. JEANNERAT. **Hita-STT, a New Five-Axis Machine Tool**. *Proceedings of the PKS Chemnitz Parallel Kinematic Seminar*, 2004. Chemnitz, Germany. 25, 26, 46, 57

[72] M. THURNEYSEN, M. SCHNYDER, R. CLAVEL, AND GIOVANOLA J. **A New Parallel Kinematics for High-Speed Machine Tools - Hita STT**. *Proceedings of the PKS Chemnitz Parallel Kinematic Seminar*, 2002. Chemnitz, Germany. 26

[73] H.-K. TÖNSHOFF. *Werkzeugmaschinen: Grundlagen*. Springer Lehrbuch, 1995. ISBN 3-540-58674-1. 1, 23, 24

[74] L.-W. TSAI. *Robot Analysis – The Mechanics of Serial and Parallel Manipulators*. Wiley-Interscience Publication, 1999. ISBN 0 471 325 937. xiv, 26

[75] M. WECK AND D. STAIMER. **Parallel kinematic machine tools – Current state and future potentials**. *CIRP Annals – Manufacturing Technology, Volume 51, Issue 2*, pages 671–683, 2002. 67

[76] Y. WU, Z. LI, H. DING, AND Y. LOU. **Quotient Kinematics Machines: Concept, Analysis and Synthesis**. *IEEE/RJS International Conference on Intelligent Robots and Systems*, 2008. Nice, France. 91

[77] W. C. YOUNG. *Roark's Formulas for Stress and Strain*. McGraw Hill, New York, 2002. 77

List of Figures

1.1	Examples of machine tools	2
1.2	Example of a complex part	4
1.3	Market survey, CECIMO Countries: Yearly production 1980-2006	5
1.4	World market survey: Global repartition of production 2005	6
1.5	Definition of the different joints	7
1.6	A serial-kinematic robot	9
1.7	A parallel-kinematic robot	10
1.8	A hybrid-kinematic robot	12
1.9	The left/right hand concept	13
1.10	Example of different DOF distributions	14
2.1	Targeted technological and economic aspects.	17
3.1	Serial and parallel kinematic mechanisms compared in terms of stiffness	20
3.2	Schematic view of a drive systems and its load	21
3.3	The free harmonic oscillator	24
3.4	The angular limits of parallel mechanisms	25
3.5	The Hita STT mechanism	26
3.6	The Alpha5 mechanism	27
3.7	The Alpha5's angular capabilities	28
4.1	Kinematics using orthogonally distributed struts	33
4.2	A collection of kinematic chains.	34
4.3	The Vision	35
4.4	The Genius 500	36
4.5	The Trijoint 900H	37
4.6	The Dumbo	37
4.7	The Mitsubishi RP	38
4.8	The Exechon	39

LIST OF FIGURES

4.9	The Sprint Z3, part of the ECOSPEED	40
4.10	The Hermes	41
4.11	The Verne	42
4.12	The Pegasus	43
4.13	The Pegasus 3-axes prototype	43
4.14	The ABB IRB 360 Flexpicker	44
4.15	The Par4 mechanism	45
4.16	The Hita STT machine tool	46
4.17	INA joints for parallel-kinematic machines	48
4.18	The Starrag Heckert SKM	50
4.19	The spherical joint of the Hita STT	51
4.20	The Tribias	52
4.21	A mixed technology joint from the EPFL.	53
4.22	The CORE joint.	54
4.23	The moving base-point drive system.	55
4.24	The variable strut-length drive system.	55
5.1	The double spherical joint of the Delta robot	59
5.2	Mobility inefficiency of different 5-axes machine tools	60
5.3	Mobility inefficiency of different 5-axes machine tools which achieve high rotation amplitudes	62
6.1	An inverted Delta and its augmented kinematics	66
6.2	Drawing of the 1st evolution spherical joint. The mechanical loop transmitting the preload is indicated.	69
6.3	1st evolution spherical joint.	69
6.4	Illustration of the point contact and definition of parameters.	70
6.5	Stiffness of the contact depending on the preload.	71
6.6	Stiffness of the contact depending on the difference of the 2 radius. . . .	72
6.7	Stiffness measuring setup	73
6.8	Stiffness of the 1st evolution spherical joint	75
6.9	Comparative study of the stiffness of different elements in a kinematic chain	76
6.10	2nd evolution of the spherical joint	77
6.11	Definition of the parameters for the modeling of the contact	78
6.12	Stamped parts.	79
6.13	Stiffness of the 2nd evolution spherical joint	80
6.14	Preloading spring induces a bending moment.	81

LIST OF FIGURES

6.15	Kinematics of the 3rd evolution of the spherical joint	82
6.16	3rd evolution of the spherical joint	83
6.17	Section through the 3rd evolution spherical joint	84
6.18	All 3 evolutions of the spherical joint. The 4th version (on the right) is the industrialized and simplified spherical joint.	85
6.19	Traction/compression and torsion load cases	87
6.20	Simple flexion load case	87
6.21	Directing efforts in traction/compression	89
6.22	Aligning efforts directly on their supports	89
6.23	Keeping force loops as short as possible	89
7.1	A 5-axes concept using twice the same module	93
7.2	Definitions for the mathematical modeling	98
7.3	Parameter optimization	103
7.4	Source-code of an optimization program	104
7.5	Numerical optimization method	105
7.6	View of the inside of a machining center.	106
7.7	Protection elements	107
7.8	The 2 different protection concepts.	108
7.9	Protection concepts for spherical/universal joints	109
7.10	Typical cabling of actuators on parallel kinematics.	111
7.11	Definition of repeatability and precision	112
7.12	Applications of the measuring cube	113
7.13	A 6-axes measuring cube	114
7.14	Measuring pure repeatability and hysteresis	114
7.15	Stiffness measuring setup	115
7.16	The frequency response of the MinAngle's left hand to a mechanical excitation	116
8.1	Design Methodology	120
8.2	Omikron5 during final detailed design	125
9.1	The Stewart3, a 3-axes orienting device	128
9.2	Available volume for the integration of the Stewart3	130
9.3	Stewart3 kinematics	131
9.4	Simple strut with 2 separated first-evolution spherical joints	132
9.5	Strut with 2 modified first-evolution spherical joints sharing a common preloading system	132

LIST OF FIGURES

9.6 Strut with 2 modified third-evolution spherical joints sharing a common preloading system 132
9.7 Strut with 2 modified third-evolution spherical joints sharing 2 identical preloading systems 133
9.8 Kinematic scheme of a ballscrew drive providing a varying strut-length . 133
9.9 The MinAngle μMachine 136
9.10 Dunlop and Orion kinematics 137
9.11 Kinematic chain of the MinAngle's left hand. 138
9.12 MinAngle kinematics 140
9.13 Comparision between a simple rotating pivot and a RCM system 141
9.14 Evolution of the pivots in the crossed bars. 142
9.15 Parasitic displacement of the MinAngle's left hand. 143
9.16 Exact division of the output angle 144
9.17 Pivot with separated, crossed blades 145
9.18 Detail pictures of the MinAngle's left hand prototype 146
9.19 Example of the MinAngle concept for machining applications 148
9.20 Omikron5: A 5-axes machine-tool 150
9.21 Example of 2 modules working on the same setup 151
9.22 Performing translations with the Omikron5 153
9.23 Pictures of the tilting head and translator of the Omikron5 154
9.24 Isostatic kinematic model of the Omikron5 154
9.25 Omikron5's flexible pivots. 155
9.26 Linear slider 157
9.27 Harmonic drive module 158
9.28 Detailed cut through the Omikron5 160

10.1 Targeted technological and economic aspects, and the contributions of this work. 163

A.2 Simplification of crossed bars system 182
A.3 Optimization of the Top and Base pivots 183
A.4 Top view of the robot output 184
A.5 Optimal pairs of L and L_x 185
A.6 Optimal RCM behaviour 186

B.1 Stiffness of joints with $\varnothing = 6mm$ and $\varnothing = 8mm$ 187
B.2 Stiffness of joints with $\varnothing = 10mm$ and $\varnothing = 12mm$ 188

LIST OF FIGURES

C.1 Definition of the geometric parameters and articular/operational coordinates of the Stewart3 190
C.2 Definition of the geometric parameters and articular/operational coordinates of the MinAngle 192
C.3 Definition of the geometric parameters and articular/operational coordinates of the Omikron5 196

D.1 2-axes double-tilting device 202
D.2 A 3-axes concept for insertion movements 203
D.3 A 3-axes concept for tilting movements. 203
D.4 A 3-axes concept for spherical movement around a RCM 204
D.5 A 3-axes concept for tilting movements. 204
D.6 A 4-axes concept for spherical movement along a translational axis . . . 205
D.7 A hybrid 5-axes concept allowing high angles of rotation. 206
D.8 A parallel 5-axes concept. 206
D.9 The Omega mechanisms 207
D.10 The 5-axes Orthoglide 207
D.11 The Metrom hybrid-kinematic machine 208
D.12 The 5-axes parallel-kinematic machine tool proposed by Feng Gao . . . 208

LIST OF FIGURES

List of Tables

1.1	A non exhaustive list of processes executed by machine tools	2
4.1	Characteristics of the INA joints, small versions	49
4.2	Characteristics of the INA joints, big versions	49
4.3	Advantages and disadvantages of flexures	51
6.1	Characteristics of the 1st evolution spherical joint	75
6.2	Characteristics of the 2nd evolution spherical joint, $\varnothing = 15mm$	81
6.3	Characteristics of the 3rd evolution spherical joint, $\varnothing = 15mm$	84
7.1	3 axes: 3 translations: **3T**	94
7.2	3 axes: **2T1R**	94
7.3	3 axes: **1T2R**	94
7.4	3 axes: **3R**	94
7.5	4 axes: 3 translations and 1 rotation: **3T1R**	95
7.6	4 axes: **2T2R**	95
7.7	4 axes: **1T3R**	95
7.8	5 axes: 3 translations and 2 rotations: **3T2R**	95
7.9	5 axes: 2 translations and 3 rotations: **2T3R**	95
7.10	6 axes: 3 translations and 3 rotations: **3T3R**	96
7.11	**Special redundant case**, 6 axes: 4 translations and 2 rotations: **4T2R**	96
8.1	Structure of the design methodology	119
9.1	Summary of the Stewart3 project requirements	129
9.2	Summary of the Stewart3's **expected** characteristics	134
9.3	Summary of the MinAngle project requirements	135
9.4	Summary of the characteristics measured on the MinAngle's left hand	146
9.5	Summary of the Omikron5 project requirements	149

LIST OF TABLES

9.6 Characteristics of flexible pivot 156
9.7 Summary of the Omikron5's measured characteristics 159

Appendix A

Optimization of the MinAngle's left hand

The main goal for the optimization is to achieve high stiffness without compromising the angular amplitude. As it is well known the stiffness of flexible hinges is in function of their amplitude, see (36). By choosing good geometrical parameters the structure perfectly distributes the angles between the pivots in a way that each pivot doesn't execute more than half of the output angle, therefore not executing high angles and guaranteeing a good stiffness.

Several optimization criteria had to be considered (see section 9.2 for the problem statement). Their evaluation is carried out in the following sections.

A.1 Crossed bars, four-bar linkage

The variation of the angles in the crossed bars system: As seen on figure A.1, the variation of the angles at the base level of the mechanism is (by symmetry):

$$variation = v_1 - \psi_1$$

At top:
$$variation = (\psi_1 + \theta) - (v_1 - \theta) = \psi_1 - v_1 + 2\theta$$

By equalizing these two variations, we assure a perfect angle distribution.

$$v_1 - \psi_1 = \psi_1 - v_1 + 2\theta \Rightarrow \psi_1 = v_1 - \theta$$

Which implies:
$$\delta = \psi_1$$

A. OPTIMIZATION OF THE MINANGLE'S LEFT HAND

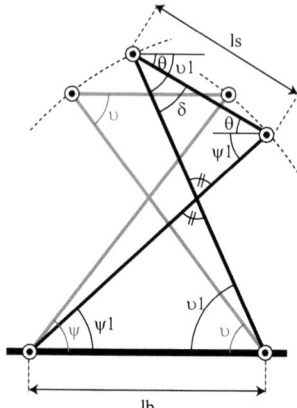

Figure A.1: Minimization of the Crossed Bars angle variation

This equation is true if and only if the two triangles formed by the crossed bars are similar. They are similar if and only if $l_b = l_s$.

In the next parts, for the sake of simplicity, the crossed bar system will be replaced by a simple revolute joint placed at the intersection of the bars see figure A.2. This can be justified because the displacement of the rotation center creates a small displacement at the output, typically 1.5 % of the length of the bars (36). It's negligible compared to the displacement due to the lever effect, around 13 % of the bar length.

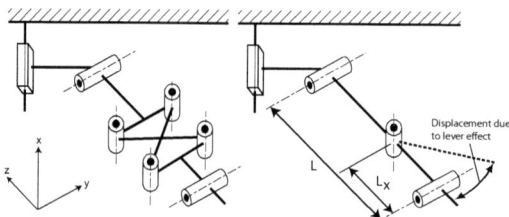

Figure A.2: Simplification of crossed bars system.

A.2 Base and Top pivots

Another point to optimize is the angular variation of the two Base and Top pivots in series. This optimization has been conducted for a particular movement i.e. keeping two actuator fixed and moving the third one in order to rotate the output platform from $\theta = -15°$ to $15°$. With such a movement only the two pivots in series (Top and Base) work in the moving arm (left arm in the figure), while in the two fixed arms (right arms in the figure) the crossed bars act like a single pivot, see figure A.3. The angles of the two serial pivot in the moving arm are studied.

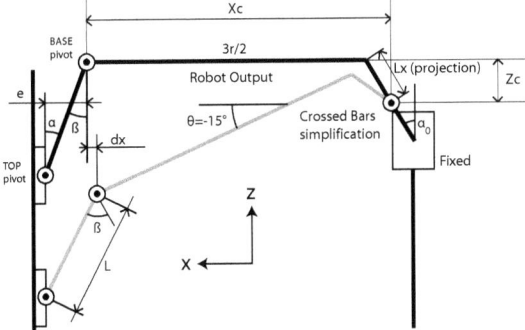

Figure A.3: Optimization of the Top and Base pivots. The plane defined by the crossed bars of the right arm is perpendicular to the plane defined by the crossed bars of the left arm. On the right side (fixed arms) only the crossed bars are moving, on the left side only the Base and Top pivots are moving.

While the robot is rotating the end of the platform fixed to the mobile arm moves along a circle centered on the crossed bars of the fixed arms. This circle lies in the X-Z plane. The displacement of the end of the platform along the Z axis doesn't change the angles in the moving arm. The displacement dx along the X axis can be computed from:

$$dx = X_c(1 - \cos\theta) + Z_c \sin\theta$$

$$dx = (\frac{3r}{2} + L_x \cos 60° \sin\alpha_0)(1 - \cos\theta) + L_x \cos\alpha \sin\theta$$

Where α_0 is the value of α when $\theta = 0$, $L_x \cos 60°$ is the projection of L_x on the X-Z plane see figure A.4. The first part of the equation is the X contribution (X_c) to dx while the second part is Z contribution (Z_c) to dx see figure A.3.

A. OPTIMIZATION OF THE MINANGLE'S LEFT HAND

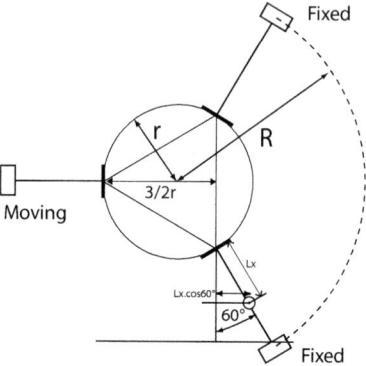

Figure A.4: Top view of the robot output.

Once dx is calculated we can derive α and β as:

$$\alpha = \arcsin \frac{e + dx}{L}$$

and:

$$\beta = \alpha - \theta$$

Then we calculate the amplitude (maximum in function of θ - minimum in function of θ) of α respectively β. The optimum is reached when the amplitude of both angles is minimum i.e. when f is minimum.

$$f = \max((\max \alpha(\theta) - \min \alpha(\theta)), (\max \beta(\theta) - \min \beta(\theta)))$$

The value of f depends on the geometric parameters: R, r, e, L, L_x. Where R is the outer radius of the machine (see figure A.4), r is the radius of the platform, L is the total length of an arm, L_x is the length from the center of the crossed bars to the end of the arm see figure A.3, e is $R - r$; all these parameters are defined in figures A.3 and A.4.

A plot of f function of L and L_x, with $R = 117, e = 30, R = 87$, is shown on figure A.5 . Chosen values for CAD design are $L_x = 70$ and $L = 140$.

A.3 Reducing the parasitic displacement

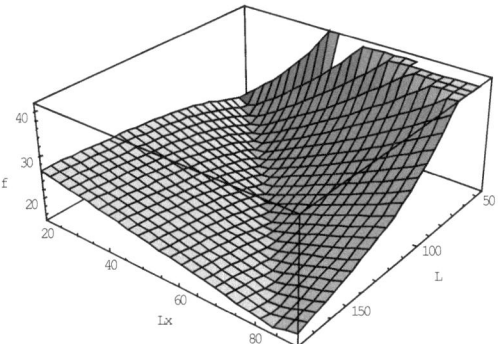

Figure A.5: f function of L and L_x. Optimal pairs of L and L_x can be found where f is minimal

A.3 Reducing the parasitic displacement

The reduction of parasitic displacement during rotation was conducted on a single geometric parameter (see figure A.6, parameter h). This parameter does not influence the other optimizations and therefore does not appear in the previous calculations.

A numerical verification of the CAD simulation showed a single optimum for the parameter h. The optimum is reached when the TCP is coincident with the center axes of the crossed bars. The resulting minimized parasitic displacement can be consulted in figure 9.15 in section 9.2.

A. OPTIMIZATION OF THE MINANGLE'S LEFT HAND

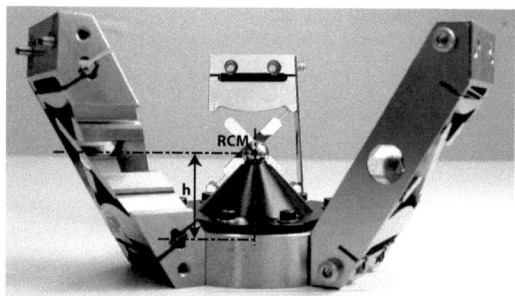

Figure A.6: Optimal RCM behaviour when the robot's output is placed at the same height as the crossed bars (their centers).

Appendix B

Stiffness of spherical joints $\varnothing = 6, 8, 10, 12mm$

This chapter presents a compilation of measurements done on spherical joints with different diameters. Besides the diameters all other geometric parameters and materials were the same as presented in section 6.2.2 for the 2nd evolution spherical joint (cone-sphere-cone). The cones and the sphere are all made of steel.

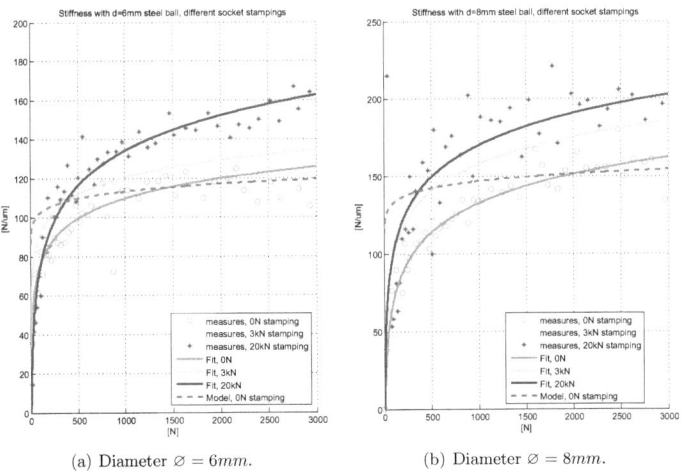

(a) Diameter $\varnothing = 6mm$. (b) Diameter $\varnothing = 8mm$.

Figure B.1: Stiffness of joints with $\varnothing = 6mm$ and $\varnothing = 8mm$.

B. STIFFNESS OF SPHERICAL JOINTS $\varnothing = 6, 8, 10, 12MM$

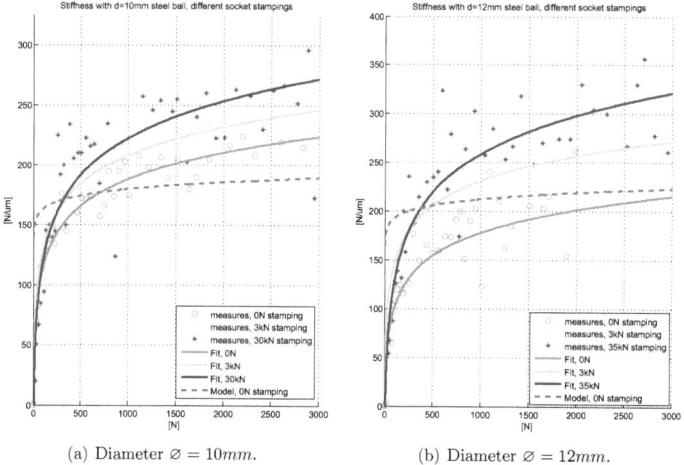

(a) Diameter $\varnothing = 10mm$. (b) Diameter $\varnothing = 12mm$.

Figure B.2: Stiffness of joints with $\varnothing = 10mm$ and $\varnothing = 12mm$.

Appendix C
Kinematic Models

Notations: Scalars are indicated as normally typeset letters (a), **vectors** as bold letters (**A**), **matrices** as bold letters with their dimensions as indices ($\mathbf{A}_{\text{nxn}}$).

The following sections provide the creation of the equation systems for all three developed prototypes. These equation system can then be solved using the numerical iterative algorithm presented in section 7.2.2.

C.1 Stewart3

The illustration of operational and articular coordinates, geometric parameters and references frames is carried out in figure C.1.

The operational coordinates **X** and the articular coordinates **Q** as well as the parasitic displacements **P** are defined as:

$$\mathbf{X} = \begin{pmatrix} \theta_x \\ \theta_y \\ z \end{pmatrix},\ \mathbf{Q} = \begin{pmatrix} q_1 \\ q_2 \\ q_3 \end{pmatrix},\ \mathbf{P} = \begin{pmatrix} x \\ y \\ \theta_z \end{pmatrix} \tag{C.1}$$

As explained in section 7.2, 2 ways to describe a same entity in the mechanism are required. In this case the points $\mathbf{A_i}$ are investigated.

One way to get $\mathbf{A_i}$ is by describing the posture of the end-effector:

$$\mathbf{A_{i'}} = \begin{pmatrix} \mathbf{R}_{3\text{x}3} & \mathbf{T}_{3\text{x}1} \\ 0 & 1 \end{pmatrix} \cdot \begin{pmatrix} a_{ix,local} \\ a_{iy,local} \\ a_{iz,local} \\ 1 \end{pmatrix} \tag{C.2}$$

C. KINEMATIC MODELS

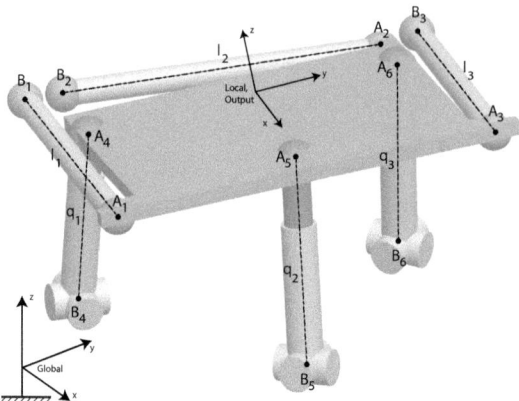

Figure C.1: Definition of the geometric parameters and articular/operational coordinates of the Stewart3 (presented in section 9.1). Points B_i are fixed, points A_i are located on the end-effector. The articular coordinates are defined as q_i and the length of the struts as l_i. The index i defines the kinematic chain. The operational coordinates are the rotations θ_x, θ_y and the vertical translation along axis z of the fixed reference frame.

where $\mathbf{R}_{3\times3}$ is the general rotation matrix defined by:

$$\mathbf{R}_{3\times3} = \mathbf{R}_z \mathbf{R}_y \mathbf{R}_x \qquad (C.3)$$

$$\mathbf{R}_{3\times3} = \begin{pmatrix} \cos\theta_z & -\sin\theta_z & 0 \\ \sin\theta_z & \cos\theta_z & 0 \\ 0 & 0 & 1 \end{pmatrix} \begin{pmatrix} \cos\theta_y & 0 & \sin\theta_y \\ 0 & 1 & 0 \\ -\sin\theta_y & 0 & \cos\theta_y \end{pmatrix} \begin{pmatrix} 1 & 0 & 0 \\ 0 & \cos\theta_x & -\sin\theta_x \\ 0 & \sin\theta_x & \cos\theta_x \end{pmatrix} \qquad (C.4)$$

and $\mathbf{T}_{3\times1}$, the translation of the mechanism's output, which is defined by:

$$\mathbf{T}_{3\times1} = \begin{pmatrix} x \\ y \\ z \end{pmatrix} \qquad (C.5)$$

The other way to obtain $\mathbf{A}_i$ is by describing the kinematic chains. As $\mathbf{A}_i$ is located on a sphere, centered in $\mathbf{B}_i$, of radius q_i or l_i, depending on the kinematic chain, following equations are obtained:

$$\mathbf{F} = \begin{pmatrix} |B_1 A_1| - l_1 \\ |B_2 A_2| - l_2 \\ |B_3 A_3| - l_3 \\ |B_4 A_4| - q_1 \\ |B_5 A_5| - q_2 \\ |B_6 A_6| - q_3 \end{pmatrix} = 0 \qquad (C.6)$$

By substituting the $\mathbf{A_i}$ in equation C.6 by the 1st description $\mathbf{A}_{i'}$ an equation system with 6 unknowns and 6 equations is obtained. The next step is to solve this system as explained in section 7.2.2 for $\mathbf{X}$ (Direct Kinematics) or $\mathbf{Q}$ (Inverse Kinematics).

C. KINEMATIC MODELS

C.2 MinAngle

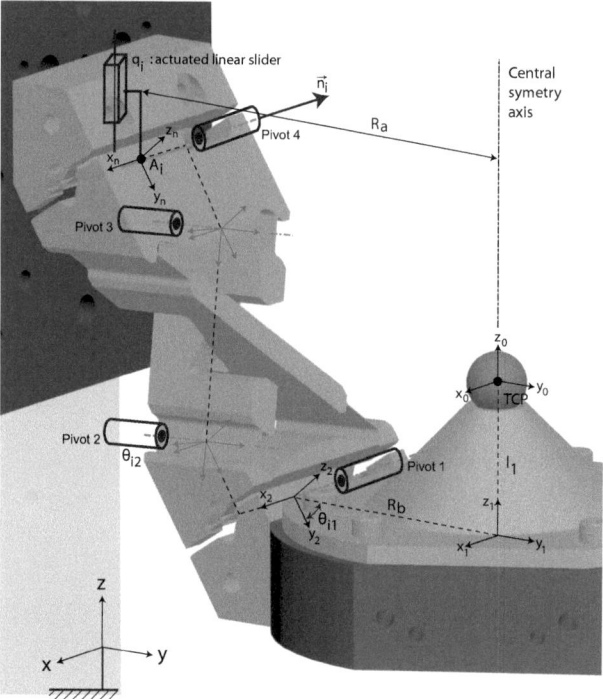

Figure C.2: Definition of the geometric parameters and articular/operational coordinates of the MinAngle (presented in section 9.2). Symbolic joints are placed as a reminder. The index i defines the kinematic chain. The operational coordinates are the rotations θ_x, θ_y and the vertical translation along z of the fixed reference frame.

The illustration of operational and articular coordinates, geometric parameters and references frames is carried out in figure C.2.

The operational coordinates $\mathbf{X}$ and the articular coordinates $\mathbf{Q}$ as well as the parasitic displacements $\mathbf{P}$ are defined:

C.2 MinAngle

$$\mathbf{X} = \begin{pmatrix} \theta_x \\ \theta_y \\ z \end{pmatrix}, \quad \mathbf{Q} = \begin{pmatrix} q_1 \\ q_2 \\ q_3 \end{pmatrix}, \quad \mathbf{P} = \begin{pmatrix} x \\ y \\ \theta_z \end{pmatrix} \quad \text{(C.7)}$$

The same global approach as for the Stewart3 will be used. Point $\mathbf{A_i}$ will be defined in 2 different ways ($\mathbf{A_{i'}}$ and $\mathbf{A_{i''}}$). The first definition comes from its location on the vertical slider. Therefore:

$$\mathbf{A_{i'}} = \begin{pmatrix} a_{ix} \\ a_{iy} \\ a_{iz} \end{pmatrix} = \begin{pmatrix} R_a \cos\alpha_i \\ R_a \sin\alpha_i \\ q_i \end{pmatrix} \quad \text{(C.8)}$$

where α_i ($0°, 120°, 240°$) is the angle defining the kinematic chains (distributed all $120°$).

The second definition of $\mathbf{A_i}$ requires a more complex approach. Starting from the TCP several homogenous matrices $\mathbf{T_{ij,4x4}}$ are provided, each of them describing one of the transformations needed to reach $\mathbf{A_i}$ along the kinematic chain (j being the index that describes the transformation, $j = 1..n$). The first transformations $\mathbf{T_{4x4}}$ are for example defined by:

$$\mathbf{T_{i1,4x4}} = \begin{pmatrix} 1 & 0 & 0 & 0 \\ 0 & 1 & 0 & 0 \\ 0 & 0 & 1 & -l_1 \\ 0 & 0 & 0 & 1 \end{pmatrix} \quad \mathbf{T_{i2,4x4}} = \begin{pmatrix} 1 & 0 & 0 & 0 \\ 0 & 1 & 0 & -R_b \\ 0 & 0 & 1 & 0 \\ 0 & 0 & 0 & 1 \end{pmatrix} \quad \text{(C.9)}$$

$$\mathbf{T_{i3,4x4}} = \begin{pmatrix} 1 & 0 & 0 & 0 \\ 0 & \cos\theta_{i1} & -\sin\theta_{i1} & 0 \\ 0 & \sin\theta_{i1} & \cos\theta_{i1} & 0 \\ 0 & 0 & 0 & 1 \end{pmatrix} \quad \text{(C.10)}$$

The matrices $\mathbf{T_{4x4}}$ which describe the rotations in *pivot 1* and *pivot 2* will introduce two additional unknowns for each kinematic chain, the angles θ_{i1} and θ_{i2}. The angle of *pivot 3* depends on the angle of *pivot 2* and therefore doesn't induce another unknown (see section A). The angle of *pivot 4* doesn't need to be known.

Point $\mathbf{A_{i''}}$, with this second approach, becomes:

$$\mathbf{A_{i''}} = \begin{pmatrix} a_{ix} \\ a_{iy} \\ a_{iz} \\ 1 \end{pmatrix} = \mathbf{T_{in,4x4}}...\mathbf{T_{i2,4x4}}\mathbf{T_{i1,4x4}} \begin{pmatrix} 0 \\ 0 \\ 0 \\ 1 \end{pmatrix} \quad \text{(C.11)}$$

C. KINEMATIC MODELS

Equalizing $A_{i'} = A_{i''}$ (equations C.8 and C.11) for each kinematic chain gives a total of 9 equations. However, because of the introduction of 2 additional unknowns for each kinematic chain (θ_{i1} and θ_{i2}) the systems comes up with a total of 12 unknowns, namely:

In case of the direct kinematics (DK): $\theta_x, \theta_y, \theta_z, x, y, z, \theta_{11}, \theta_{12}, \theta_{21}, \theta_{22}, \theta_{31}, \theta_{32}$

In case of the inverse kinematics (IK): $q_1, q_2, q_3, x, y, \theta_z, \theta_{11}, \theta_{12}, \theta_{21}, \theta_{22}, \theta_{31}, \theta_{32}$

The information which was not yet implemented as an equation is the collinearity condition of the axis of *pivot 4*. In fact, until now, we just calculated its position. This will provide the missing 3 equations to complete our equation system.

The axis of the pivot is defined by the geometry of the robot. Since the robot has a 120° symmetry:

$$\mathbf{n_i} = \begin{pmatrix} -R_a sin\alpha_i \\ R_a cos\alpha_i \\ 0 \end{pmatrix} \quad (C.12)$$

The y-axis $\mathbf{y_{in}}$ of the last reference frame is then computed. The y-axis should be perpendicular to $\mathbf{n_i}$.

$$\mathbf{y_{in}} = \mathbf{T_{in,4x4}} ... \mathbf{T_{i2,4x4}} \mathbf{T_{i1,4x4}} \begin{pmatrix} 0 \\ 1 \\ 0 \end{pmatrix} \quad (C.13)$$

The perpendicularity is verified by the scalar product:

$$\mathbf{n_i} \cdot \mathbf{y_{in}} = 0 \quad (C.14)$$

This provides the last 3 equations in order to complete our equation system:

C.2 MinAngle

$$\mathbf{F} = \begin{pmatrix} \begin{pmatrix} R_a cos\alpha_1 \\ R_a sin\alpha_1 \\ q_1 \\ 1 \end{pmatrix} - \mathbf{T_{1n,4x4}}...\mathbf{T_{12,4x4}}\mathbf{T_{11,4x4}} \begin{pmatrix} 0 \\ 0 \\ 0 \\ 1 \end{pmatrix} \\ \begin{pmatrix} R_a cos\alpha_2 \\ R_a sin\alpha_2 \\ q_2 \\ 1 \end{pmatrix} - \mathbf{T_{2n,4x4}}...\mathbf{T_{22,4x4}}\mathbf{T_{21,4x4}} \begin{pmatrix} 0 \\ 0 \\ 0 \\ 1 \end{pmatrix} \\ \begin{pmatrix} R_a cos\alpha_3 \\ R_a sin\alpha_3 \\ q_3 \\ 1 \end{pmatrix} - \mathbf{T_{3n,4x4}}...\mathbf{T_{32,4x4}}\mathbf{T_{31,4x4}} \begin{pmatrix} 0 \\ 0 \\ 0 \\ 1 \end{pmatrix} \\ \mathbf{n_1} \cdot \mathbf{y_{1n}} \\ \mathbf{n_2} \cdot \mathbf{y_{2n}} \\ \mathbf{n_3} \cdot \mathbf{y_{3n}} \end{pmatrix} = 0 \qquad (C.15)$$

C. KINEMATIC MODELS

C.3 Omikron5

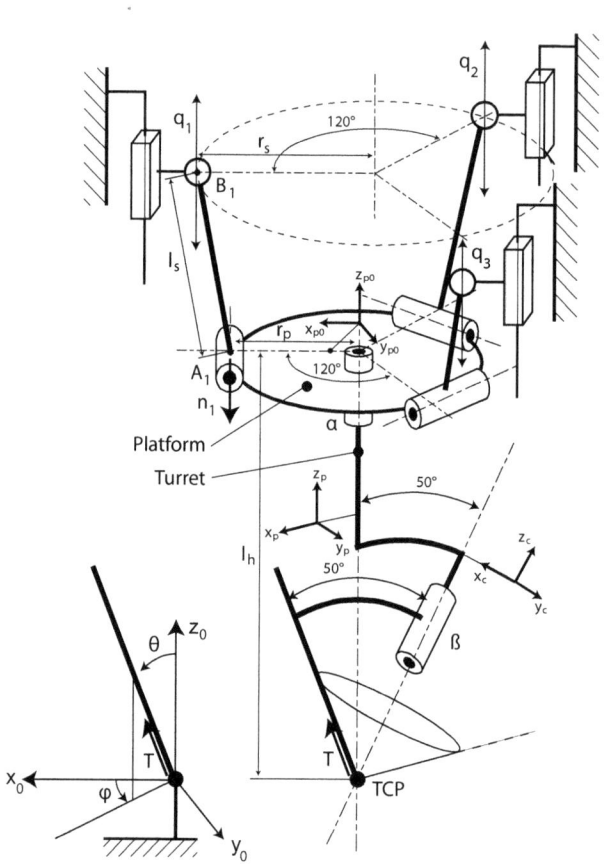

Figure C.3: Definition of the geometric parameters and articular/operational coordinates of the Omikron5 (presented in section 9.3). The figure illustrates the mechanism in a general pose. Points A_i are located in the centers of the spherical joints. Points B_i are located in the intersection of the pivot's axes and the plane that is perpendicular to the pivot's axes and which is going through the center of the platform. The index $i = 0..2$ indicates the kinematic chain. The vertical sliders and pivots on the platform are distributed symmetrically all 120°. The vector **T** symbolizes the tool.

The illustration of operational and articular coordinates, geometric parameters and references frames is carried out in figure C.3.

The operational coordinates $\mathbf{X}$ and the articular coordinates $\mathbf{Q}$ are defined as:

$$\mathbf{X} = \begin{pmatrix} x \\ y \\ z \\ \theta \\ \phi \end{pmatrix}, \quad \mathbf{Q} = \begin{pmatrix} q_1 \\ q_2 \\ q_3 \\ \alpha \\ \beta \end{pmatrix} \tag{C.16}$$

The operational coordinate θ is defined as the angle between the vertical axis z_0 (of fixed, global references system R_0) and the tool. The other rotative operation coordinate, ϕ, is defined as the angle between x_0 and the projection of the tool in the plane defined by x_0 and y_0.

In a first phase, the equations for the *translations of the output* will be derived. The rotation of the end-effector around a RCM allows to decouple the translations from the rotations (the inverse, in this case, is not possible). Translating the end-effector generates rotations in the platform (see figure 9.22 in section 9.3). Therefore, 3 additional unknowns a, b, c are introduced. They describe the rotations of the platform around the fixed, global reference system $R_0 : (x_0, y_0, z_0)$:

$$\mathbf{W} = \begin{pmatrix} a \\ b \\ c \end{pmatrix} \tag{C.17}$$

In order to create the equation system two descriptions for the points $\mathbf{A_i}$ ($\mathbf{A_{i'}}$ and $\mathbf{A_{i''}}$) will be carried out. Additionally, since these points are located on pivots, a collinearity condition for the pivot axes will be defined for both approaches.

The first description of $\mathbf{A_i}$ is provided by the end-effector posture:

$$\mathbf{A_{i'}} = \begin{pmatrix} \mathbf{R}_{3x3} & \mathbf{T}_{3x1} \\ 0 & 1 \end{pmatrix} \cdot \begin{pmatrix} r_p cos(i\frac{2\pi}{3}) \\ r_p sin(i\frac{2\pi}{3}) \\ l_n \\ 1 \end{pmatrix} \tag{C.18}$$

Where $\mathbf{R}_{3x3} = \mathbf{R_{z0}R_{y0}R_{x0}}$ is the general rotation matrix around the axes of $R_0 : (x_0, y_0, z_0)$ with the angles given for $\mathbf{W} = (a, b, c)^T$, and $\mathbf{T}_{3x1} = (x, y, z)^T$ is the translation vector containing the translational, operational coordinates.

The second description is provided by writing down the equation of the sphere located in $\mathbf{B_i}$ (with radius l_s):

C. KINEMATIC MODELS

$$|\mathbf{B_l A_{l''}}| - l_s = 0 \quad \text{(C.19)}$$

where $\mathbf{B_l}$ contains the translational, articular coordinates and is defined by:

$$\mathbf{B_l} = \begin{pmatrix} r_s cos(i\frac{2\pi}{3}) \\ r_s sin(i\frac{2\pi}{3}) \\ q_i \end{pmatrix} \quad \text{(C.20)}$$

Substituting $\mathbf{A_{l''}}$ by $\mathbf{A_{l'}}$ in equation C.19 results in the first 3 equations:

$$\begin{pmatrix} |\mathbf{B_1 A_{1'}}| - l_s \\ |\mathbf{B_2 A_{2'}}| - l_s \\ |\mathbf{B_3 A_{3'}}| - l_s \end{pmatrix} = 0 \quad \text{(C.21)}$$

The 3 additional, necessary equations are provided by a perpendicularity condition[1] of the pivot's axis (in point $\mathbf{A_l}$) and the strut segment $\mathbf{A_l B_l}$. In order to do this, a description of the joint's axis is first needed:

$$\mathbf{n_l} = \begin{pmatrix} \mathbf{R_{3x3}} & \mathbf{T_{3x1}} \\ 0 & 1 \end{pmatrix} \cdot \begin{pmatrix} -sin(i\frac{2\pi}{3}) \\ cos(i\frac{2\pi}{3}) \\ 0 \\ 1 \end{pmatrix} \quad \text{(C.22)}$$

The perpendicularity condition provides the 3 following equations:

$$\begin{pmatrix} \mathbf{n_1} \cdot \mathbf{A_{1'} B_1} \\ \mathbf{n_2} \cdot \mathbf{A_{2'} B_2} \\ \mathbf{n_3} \cdot \mathbf{A_{3'} B_3} \end{pmatrix} = 0 \quad \text{(C.23)}$$

Equations C.21 and C.23 provide the 6 equations needed to solve the IK (unknowns: q_1, q_2, q_3, a, b, c) and DK (unknowns: x, y, z, a, b, c) for translations.

When considering *rotations* the equation system has to be augmented by 2 equations because of 2 additional articular and operational coordinates. In order to model the serial-kinematic head several reference systems are introduced, in addition to the fixed, global reference system R_0:

$R_{p0} : (x_{p0}, y_{p0}, z_{p0})$: Fixed to the platform and rotated, with respect to R_0, by the angles a, b, c. These angles are induced by the parallel-kinematic translator.

$R_p : (x_p, y_p, z_p)$: Fixed to the turret. This reference system is rotated, with respect to R_{p0}, by the angle α (articular coordinate) around the common axis $z_p = z_{p0}$.

---
[1]Which is mathematically the same as describing the collinearity of two axes.

C.3 Omikron5

$R_c : (x_c, y_c, z_c)$: Fixed to the turret, with z_c aligned to the β-pivot. R_c is rotated, with respect to R_p, by an angle of -50° around the common axis $y_p = y_c$.

Following transformations can be stated:

$$R_0 \rightarrow R_{p0} : \begin{pmatrix} x_{p0} \\ y_{p0} \\ z_{p0} \end{pmatrix} = \mathbf{R}_{3x3} \begin{pmatrix} x_0 \\ y_0 \\ z_0 \end{pmatrix} \tag{C.24}$$

$$R_{p0} \rightarrow R_p : \begin{pmatrix} x_p \\ y_p \\ z_p \end{pmatrix} = \begin{pmatrix} cos\alpha & -sin\alpha & 0 \\ sin\alpha & cos\alpha & 0 \\ 0 & 0 & 1 \end{pmatrix} \begin{pmatrix} x_{p0} \\ y_{p0} \\ z_{p0} \end{pmatrix} = \mathbf{M}_{1,3x3} \begin{pmatrix} x_{p0} \\ y_{p0} \\ z_{p0} \end{pmatrix} \tag{C.25}$$

$$R_p \rightarrow R_c : \begin{pmatrix} x_c \\ y_c \\ z_c \end{pmatrix} = \begin{pmatrix} cos50° & 0 & -sin50° \\ 0 & 1 & 0 \\ sin50° & 0 & cos50° \end{pmatrix} \begin{pmatrix} x_p \\ y_p \\ z_p \end{pmatrix} = \mathbf{M}_{2,3x3} \begin{pmatrix} x_p \\ y_p \\ z_p \end{pmatrix} \tag{C.26}$$

Then the tool vector $\mathbf{T}$, which lies on a cone with an aperture angle of 50°, is specified with respect to R_c:

$$\mathbf{T}_{R_c} = \begin{pmatrix} cos\beta \\ sin\beta \\ \frac{1}{tan50°} \end{pmatrix} \tag{C.27}$$

And in normalized form:

$$\mathbf{T}_{norm,R_c} = \begin{pmatrix} sin50° cos\beta \\ sin50° sin\beta \\ cos50° \end{pmatrix} \tag{C.28}$$

The tool vector $\mathbf{T}_{norm,R_c}$ is then transferred in the fixed, global reference system R_0:

$$\mathbf{T}_{norm,R_0} = \begin{pmatrix} x_T \\ y_T \\ z_T \end{pmatrix}_{R_0} = \mathbf{R}_{3x3} \mathbf{M}_{1,3x3} \mathbf{M}_{2,3x3} \mathbf{T}_{norm,R_c} \tag{C.29}$$

The operational coordinates can now be deduced from the tool vector:

$$\theta = arccos(z_t) \tag{C.30}$$

$$\phi = arccos\left(\frac{x_T}{\sqrt{(x_T)^2 + (y_T)^2}}\right) \tag{C.31}$$

These 2 equations are added to the existing 6 (equations C.21 and C.23) and result in the complete, final system of 8 equations and 8 unknowns. Depending on the model we have the following unknown variables:

C. KINEMATIC MODELS

In case of the direct kinematics (**DK**): $x, y, z, \theta, \phi, a, b, c$

In case of the inverse kinematics (**IK**): $q_1, q_2, q_3, \alpha, \beta, a, b, c$

Appendix D
Catalogue of Kinematics

The aim of this appendix is to compile a catalogue with new or existing kinematics which were not illustrated in the previous chapters. The first section suggests some completely new kinematics, whereas the second section illustrates existing kinematics. The existing kinematics are illustrated for the sake of completeness, as they are used as references in the main document.

D. CATALOGUE OF KINEMATICS

D.1 New Kinematics

D.1.1 2-axes

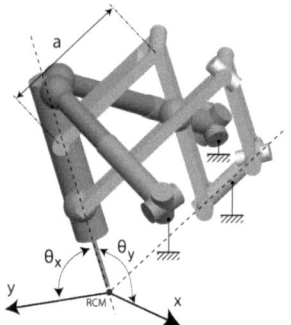 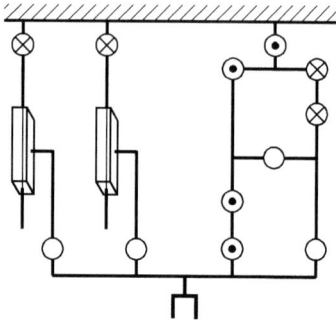

(a) A 2-axes double-tilting device which is always pointing towards a RCM. The placement of the RCM can be chosen by varying the width of the parallelogram defined by a.

(b) Kinematics of the 2-axes concept. The isostatic system presents a total sum of mobilities of 26. However, by using only pivots in the right part of the kinematics we can greatly reduce this amount. Though, the mechanical segments must then be designed to allow a localized and controlled deformation in order to prevent the system of being overconstrained.

Figure D.1: A 2-axes double-tilting device generating rotations around a RCM (θ_x, θ_y). Mobility inefficiency is 13.

D.1.2 3-axes

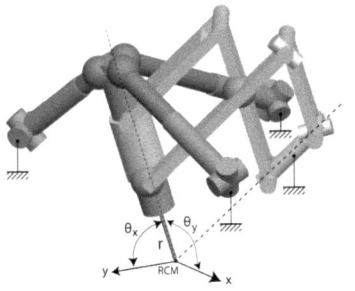
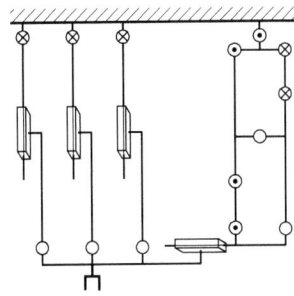

(a) A 3-axes concept for insertion movements (spherical coordinates r, θ_x, θ_y) with RCM capabilities.

(b) Kinematics of the concept.

Figure D.2: A 3-axes concept for insertion movements. The mechanism is an evolution of the 2-axes concept present in figure D.1. Mobility inefficiency is 11.

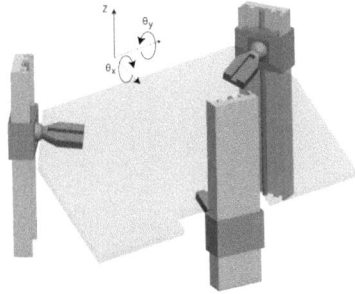
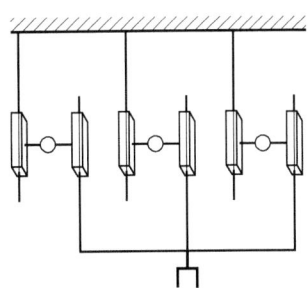

(a) A 3-axes mechanism for tilting movements. The mechanism allows movements around θ_x, θ_y and along Z.

(b) The stroke of the sliders placed on the output can be kept small, depending on the targeted rotation amplitudes. The integration of flexible joints for these sliders is imaginable.

Figure D.3: A 3-axes concept with the same mobilities as the mechanism in figure D.5. The presented kinematics is a variant of Scussat's Sixtiff (67). The 3 kinematic chains are all identical. The mobility inefficiency is 5.

D. CATALOGUE OF KINEMATICS

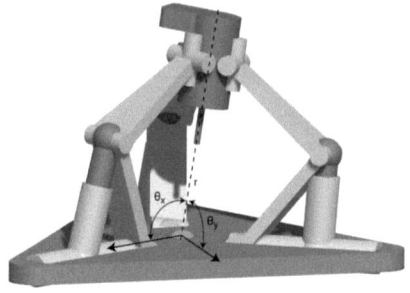
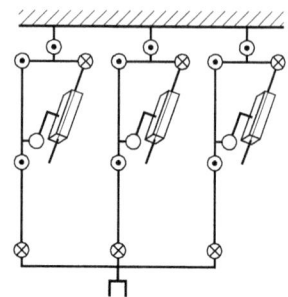

(a) A 3-axes concept for insertion movements (spherical coordinates r, θ_x, θ_y) with RCM capabilities. The tool is always aligned with the central workpiece (symbolized by the central sphere).

(b) Kinematics of the 3-axes insertion mechanism.

Figure D.4: A 3-axes concept for spherical movement around a RCM. The system presents a total sum of mobilities of 33 which leads to a mobility inefficiency of 11.

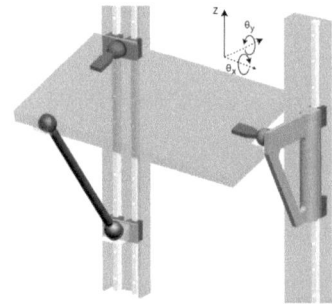

(a) A 3-axes mechanism for tilting movements. The mechanism allows movements around θ_x, θ_y and along Z.

(b) Kinematics of the 3-axes "point-line-plane" mechanism.

Figure D.5: A 3-axes concept for tilting movements and 1 translation, the "point-line-plane" mechanism. In the proposed mechanism 2 axes are sharing the same linear bearing, allowing by this an effective and room-saving design. The system presents a total sum of 15 mobilities which leads to a mobility inefficiency of 5.

D.1 New Kinematics

D.1.3 4-axes

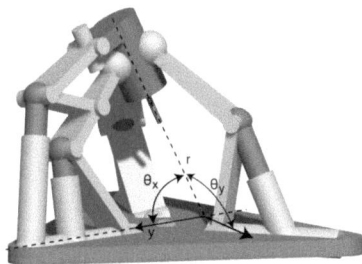 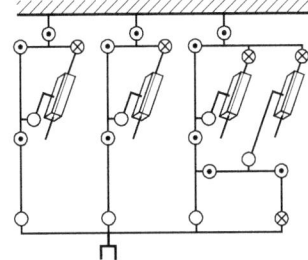

(a) A 4-axes module for insertion movements (spherical coordinates r, θ_x, θ_y and translation along y). The tool is always aligned on the y-axis.

(b) Kinematics of the 4-axes mechanism. The 4th axis, represented on the right, is designed to allow high rotation angles, using a four-bar-linkage (like on excavators).

Figure D.6: A 4-axes concept for spherical movement along a translational axis. An evolution of the 3 axes concept presented in figure D.4. The system presents a total sum of mobilities of 46 which leads to a mobility inefficiency of 11.5.

D. CATALOGUE OF KINEMATICS

D.1.4 5-axes

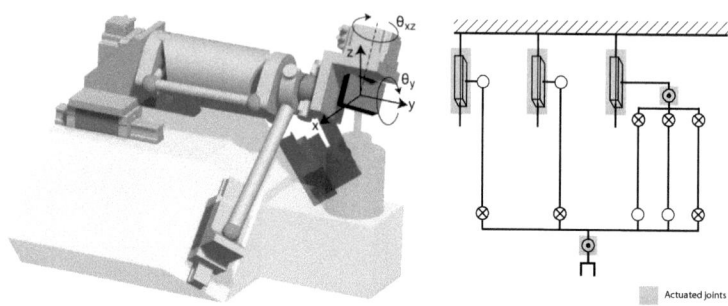

(a) A 5-axes concept with movements along x, y, z, θ_y and θ_{xz}.

(b) Kinematic of the 5-axes concept.

Figure D.7: A hybrid 5-axes concept allowing high angles of rotation. This concept is well suited if all 5 DOF need to be placed on 1 hand. It presents a total sum of mobilities of 29 which lead to a mobility inefficiency of 5.8.

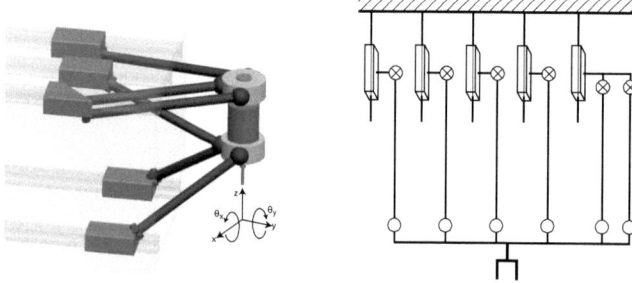

(a) A 5-axes concept with movements along x, y, z, θ_y and θ_x.

(b) Kinematic of the 5-axes concept.

Figure D.8: A parallel 5-axes concept. A parallelogram eliminates the 6th mobility. Mobility inefficiency is 7.

D.2 Additional, Existing Kinematics

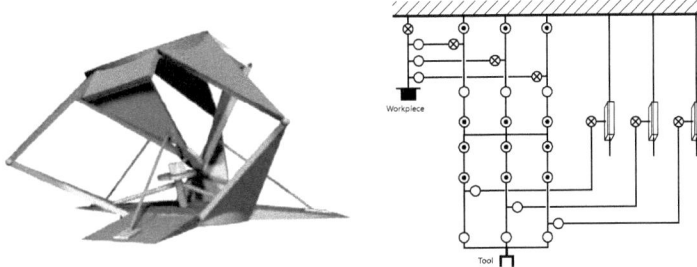

(a) The Omega mechanism.

(b) Isostatic kinematic scheme of the Omega mechanism.

Figure D.9: The fully-parallel 5-axes Omega (18) $(x, y, z, \theta_x = \pm 90°, \theta_y = \pm 90°)$.

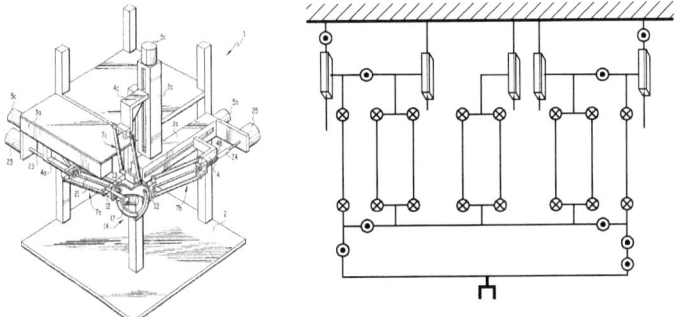

(a) The 5-axes Orthoglide.

(b) Kinematic scheme of the 5-axes Orthoglide. As it is represented in the patent it is 20 times overconstrained.

Figure D.10: The fully-parallel 5-axes Orthoglide (13) $(x, y, z, \theta_x = \pm 90°, \theta_y = \pm 90°)$.

D. CATALOGUE OF KINEMATICS

(a) The Metrom parallel kinematic machine. (www.metrom.com)

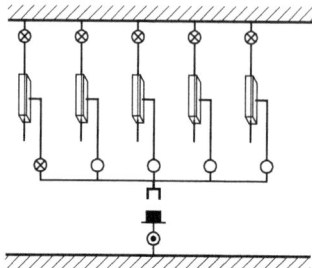
(b) Kinematic scheme of the Metrom parallel kinematic machine. The machine possesses 1 redundant actuator.

Figure D.11: The hybrid, redundant 5-axes Metrom parallel-kinematic machine (www.metrom.com, $x, y, z, \theta_x = \pm 45°, \theta_y = \pm 180°$).

(a) The 5-axes parallel-kinematic machine tool proposed by Feng Gao.

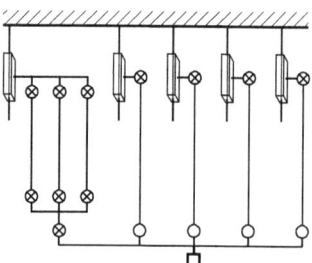
(b) Kinematic scheme. The mechanism, as it is realized, is 2 times overconstrained. From the 3 struts on the extreme left 2 should contain 1 spherical joint instead of a universal joint in order to be isostatic.

Figure D.12: The fully-parallel machine tool proposed by Feng Gao (27) ($x, y, z, \theta_x = \pm 45°, \theta_y = \pm 45°$).

Die VDM Verlagsservicegesellschaft sucht für wissenschaftliche Verlage abgeschlossene und herausragende

Dissertationen, Habilitationen, Diplomarbeiten, Master Theses, Magisterarbeiten usw.

für die kostenlose Publikation als Fachbuch.

Sie verfügen über eine Arbeit, die hohen inhaltlichen und formalen Ansprüchen genügt, und haben Interesse an einer honorarvergüteten Publikation?

Dann senden Sie bitte erste Informationen über sich und Ihre Arbeit per Email an *info@vdm-vsg.de*.

Sie erhalten kurzfristig unser Feedback!

VDM Verlagsservicegesellschaft mbH
Dudweiler Landstr. 99
D - 66123 Saarbrücken
www.vdm-vsg.de

Telefon +49 681 3720 174
Fax   +49 681 3720 1749

Die VDM Verlagsservicegesellschaft mbH vertritt

Printed by Books on Demand GmbH, Norderstedt / Germany